MW00835332

Multilingual Book Production

Multilingual Book Production

Edited by Bill Cope and Gus Gollings

C O M M O N
G R O U N D

C-2-C Project: Book 2.2

Technology Drivers Across the Book Production
Supply Chain, From the Creator to the Consumer

This book is published at theHumanities.com and C-2-CProject.com
Series imprints of theUniversityPress.com

First Published in Australia in 2001
By Common Ground Publishing Pty Ltd
PO Box 463
Altona Vic 3018
ABN 66 074 822 629
www.theHumanities.com
www.C-2-CProject.com

National Library of Australian Cataloguing-in-Publication data:

    C-2-C project: technology drivers across the book
    production supply chain, from creator to consumer. Book 2.2,
    Multilingual book producation.

ISBN 1 86335 073 X (pbk.).

ISBN 1 86335 074 8 (PDF).

1. Publishers and publishing. 2. Printing industry –
Technological innovations. I. Cope, Bill, 1957– . II.
Gollings, Gus. III. Title: Multilingual book production.

070.5

Cover designed by Diana Kalantzis.
Typeset in Garamond and Optima by Common Ground Publishing.
Printed in Australia by dbooks on 79gsm Bulky Paperback.

FOREWORD

This series of four books examines technology drivers in the book production supply chain. The series represents the latest stage of a project that began in 2000 examining book production in the digital age. It is part of a larger research project funded by Australia's Department of Industry, Science and Resources.

The theme of the series is encapsulated in the project's name – C-2-C – or Creator to Consumer. In one respect, this name reflects a persistent theme about the relationship between commerce and the new information and communication technologies, and it reflects this through an ironical play on that overly used expression, B-2-B. Online business-to-business transactions, it seems, are today's technological key to greater business efficiency and enhanced competitiveness. The principal effect of B-2-B developments, we are told, will be disintermediation of the supply chain. Some old players will disappear, others will transform themselves, and new ones will emerge. And, whoever the players prove to be, and that is by no means clear, it is clear that complementary businesses will form even closer relationships. The authors' irony notwithstanding, B-2-B developments will certainly have many of these effects.

However, the B-2-B idea all too often represents a limited view of the supply chain. The idea of B-2-C, or business to consumer, tacks human needs and interests onto the end of the supply chain as a kind of afterthought, as a subset of retailing. And the creative end of the process – conception, design, knowledge, research, culture; the very reasons why there is something to supply in the first place – is often almost entirely overlooked. Hence the C-2-C notion introduced by the authors of these books. This notion is as concerned with the creation and consumer ends of the digital world as it is with the B-2-B interaction in the middle. Indeed, without creativity and use as the end points in the process, the exchange links along the chain are empty commercial calculations and mere technological tricks without human purpose.

This is a series of books about books. It describes the relationships of new technologies to the daily lives and changing work practices of players all along the book production supply chain, from creators (authors, in the case of books), to publishers, to printers, to bookstores and finally to consumers. It is about changes in commerce, in technology and in culture – culture understood here as the way we live, as well as the culture of business itself as exemplified by the creation, manufacture and dissemination of cultural contents through the book trade.

Needless to say, everything in the book trade is changing. These changes are very much influenced by the technologies of the Internet, digital printing, etext reading devices, digital rights management, digital design tools, digital content capture and archiving systems, and the many other elements of technological change which this series of books describes in vast detail.

As it transpires, these are not just books about books. They also represent a considered reflection on the interlinkage of broader social changes, which are always simultaneously technological, commercial and cultural.

They are also telling statements of the strengths of RMIT University and the Faculties we lead. We are proud to be able to say we are a technology university, and much more. For, to be the outstanding technology university we are, we have to be as concerned with culture (creation and use) as we are with commerce and its technological means.

This is why the research upon which this series of books is based has of necessity been an interdisciplinary endeavour, drawing on the expertise of academics of diverse disciplinary backgrounds from right across the University.

We live at what will in all probability in the future be seen to be a defining moment in human history. Whether that moment is seen primarily as a technological moment (the digital era), or a commercial moment (a high point for market capitalism), or a cultural moment (of globalised cultural diversity), remains to be seen. We do believe, however, that these books, in their own meticulous way, document some important aspects of that moment.

Prof. Robin Williams
Dean, Faculty of Art, Design and Communication
RMIT University

Prof. Mary Kalantzis
Dean, Faculty of Education, Language and Community Services
RMIT University

Contents

C-2-C Project Advisory Committee

Philip Anderson, Print Industries Association of Australia;
Patrick Bernau, Fuji Xerox Australia;
José Borghino, Australian Society of Authors;
Susan Bridge, Australian Publishers Association;
Tony Burch, Consultant; Michael Fraser, Copyright Agency Limited;
Garry Knespal, Graphic Arts Services Association of Australia;
Kathleen Mapperson, Australian Booksellers Association;
Anni Rowland-Campbell, Graphic Arts Merchants Association of Australia;
Ivan Trundle, Australian Library and Information Association;
Richard Vines, Enhanced Printing Industry Competitiveness Scheme/Print Industries Association of Australia;
Michael Webster, RMIT University, Copyright Agency Limited, Booktrack; Dianne Woodward, Australian Campus Booksellers Association

NOIE Representative
Luke Naismith, National Office of the Information Economy.

EPICS Representative
Jason Ashurst, Enhanced Printing Industry Competitiveness Scheme

C-2-C Project Team

Billy Atta, Teacher, School of Printing, RMIT University;
Stuart Bishop, Software Engineer, Common Ground;
Robert Black, Manager, Enterprise Training and International Projects, RMIT University;
Melissa Brown, Research Assistant, Globalisation & Cultural Diversity Research Concentration, RMIT University;
Peter Burrows, Telstra Home Team, Interactive Information Institute, RMIT University;
Michael Coburn, Telstra Home Team, Interactive Information Institute, RMIT University;
Bill Cope, Director, Common Ground;
Rob Dunn, Teacher, School of Printing, RMIT University;
Robin Freeman, Publisher, Common Ground;
Laurie Gerber, Language Technology Broker;

Gus Gollings, C-2-C Project Manager, Common Ground;

James Hawthorne, Editor, Common Ground;

Ray Hester, Teacher, School of Printing, RMIT University;

Diana Kalantzis, Director, Common Ground;

Terry Laidler, Director of the Centre for International Research in Communication & Information Technologies, RMIT University;

Daria Loi, Lecturer, School of Architecture and Design & Telstra Home Team, RMIT University;

John Magnik, Teacher, School of Printing, RMIT University;

Dean Mason, Business Development Manager, Common Ground;

Rod Palmer, Senior Software Engineer, Common Ground;

David Prater, Research Assistant, Centre for International Research in Communication & Information Technologies, RMIT University;

Andrew Readman, Teacher, School of Printing, RMIT University;

Michael Singh, Professor of Language and Culture, RMIT University;

Atsushi Takagi, Program Coordinator, Applied Language, Department of Language & International Studies, RMIT University;

Stavroula Tsembas, Education Technology Manager, Faculty of English, Language & Community Services, RMIT University;

Linda Wilkins, Telstra Home Team, Interactive Information Institute, RMIT University;

Christopher Ziguras, Post Doctoral Fellow, Department of Languages & International Studies, RMIT University.

AN INTRODUCTION TO THE EPICS GRANTS PROGRAM

Support for Australia's Book Production Industry

Australia's book production industry is going through significant changes. Rapid advances in technology are resulting in major changes to the way books are designed, manufactured, and distributed. Australian companies can reap the benefits of these changes by adopting new ways of doing business to enhance their competitiveness in the global market.

The Commonwealth Government has developed a $240 million package of measures known as the Book Industry Assistance Plan to provide support for printing, publishing, book selling, book authorship and library activities in Australia. From this amount, $48 million is to be provided over four years for Book Production Enhanced Printing Industry Competitiveness Scheme (EPICS) grants.

The Book Production-EPICS Grants, which commenced on 1 July 2000, aim to improve the development and competitiveness of the Australian book production industry. The program provides grants to industry for projects that encourage innovation, business development and skills formation consistent with the implementation of PRINT21- the Printing Industries Action Agenda.

EPICS Program Objectives

The objectives of the Book Production-EPICS Grants program are to:

- enhance the competitiveness of firms involved in the Australian book production industry through encouraging innovation, infrastructure development, business development, training and skills formation;
- increase industry linkages and best practice dissemination;
- improve the capacity of the industry to satisfy consumer demand; and
- improve the book production industry's contribution to the national economy, which may increase employment opportunities in regional Australia.

Key Elements of Book Production-EPICS Grants

The key elements of Book Production-EPICS Grants are:

- an Enterprise Development Fund (EDF), which is administered by AusIndustry and funds proposals from individual companies or small groups of applicants with commercial objectives aimed at increasing their competitiveness;
- an Infrastructure and Industry Growth Fund (IIGF), which is administered by the Manufacturing Engineering and Construction (MEC) Division of the Department of Industry Science and Resources and funds proposals aimed at delivering industry-wide benefits; and
- a professional Client Manager Service which is administered by MEC, together with the Printing Industries Association of Australia (PIAA), and assists firms to access the Book Production-EPICS Grants as well as provide referrals to other government industry assistance programs and services.

In general, if there is commercial benefit to be generated from the project at the company level, EDF is the most appropriate fund. If the balance of benefits or the demonstration potential of the project is more applicable to the industry as a whole, IIGF is the most appropriate fund.

Important Trends Within the Australian Book Production Industry

The Australian Government's establishment of the EPICS scheme is a timely development for the Australian book industry, because the industry is at the early stages of responding to:

- the necessity of adopting core competencies underpinned by digital asset management and real time digital rights management activities;
- emerging customer demands associated with mass customisation, multi-channel publishing and distribution;
- the need to integrate risk management strategies (that are becoming available by the uptake of print on demand technologies within the publishing and printing industries' business models);
- the benefits available via the adoption of XML repository systems and supply chain enhancement via digital workflow systems.

Enterprise Development

The following model has been prepared to summarize, in part, some of the tactical strategies that publishers / printers might need to consider in unlocking the possibilities associated with digital asset management and the integration of print on demand (POD) and mass customisation competencies into their business models.

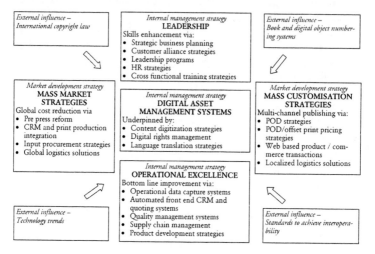

The model highlights the fact that the boundaries between traditional publishing and printing activities are likely to become increasingly blurred and that underpinning all activities is the need for the entire supply chain to become literate in digital content and rights management practices.

I am sure that this collaboration between Common Ground and RMIT will contribute to the research base to enable both publishers and printers to select and implement appropriate strategies to secure enhanced competitiveness and profitable business growth outcomes that will benefit the Australian book production industry.

Richard Vines
EPICS Client Manager, August 2001

ACKNOWLEDGEMENTS

We acknowledge the support and funding of the Australian Federal Government through the Commonwealth Department of Industry, Science and Resources, the Infrastructure and Industry Growth Fund (IIGF), Book Production Enhanced Printing Industry Competitiveness Scheme (EPICS) Grants. Appreciation is expressed in particular to the Print and Paper Industries Section, Manufacturing, Engineering and Construction Division.

We acknowledge also the dedicated team of industry professionals who graciously donated their time as members of the C-2-C Project Advisory Committee: Philip Andersen, Patrick Bernau, Jose Borghino, Susan Bridge, Tony Burch, Michael Fraser, Garry Knespal, Kathleen Mapperson, Anni Rowland-Campbell, Ivan Trundle, Richard Vines, Michael Webster and Dianne Woodward. Appreciation also to Luke Naismith for representing the National Office of the Information Economy (NOIE) as an observer.

This series of research documents has been created through a team effort incorporating the labours of many people. Within RMIT University there has been a wide cross-section of departments and people involved from the project's inception. Especially, we would like to thank Billy Atta, Robert Black, Melissa Brown, Peter Burrows, Lynda Caldwell, Michael Coburn, Bob Dunn, Ray Hester, Mary Kalantzis, Terry Laidler, Daria Loi, John Magnik, Peter Phipps, Sue Pickles, David Prater, Andrew Readman, Ian Sapwell, Michael Singh, Atsushi Takagi, Stavroula Tsembas, Linda Wilkins, Robin Williams and Christopher Ziguras.

Everyone at Common Ground has directly or indirectly helped bring this work to completion. Appreciation is expressed for their support to: Stuart Bishop, Bill Cope, Katherine Czerwinski, Robin Freeman, Gus Gollings, James Hawthorne, Diana Kalantzis, Dean Mason, Kathryn Otte, Rod Palmer and Irene Wong.

ABOUT THE CONTRIBUTORS

Melissa Brown teaches in the RMIT language program and is a researcher with the RMIT Globalisation and Cultural Diversity Research Concentration. She has worked at the Asia Pacific regional Office of UNESCO, Bangkok, researching and writing about educational development projects in the Asia Pacific region. Her research interests include material culture of post WWII migration and material expressions of culture and ethnicity.

Dr Bill Cope has been a Director of Common Ground Publishing since 1984, and is currently working on experimental mixed-medium print and Internet publishing, www.C-2-CSystem.com. He is a former First Assistant Secretary in the Department of the Prime Minister and Cabinet and director of the Office of Multicultural Affairs. His academic research and writing crosses a number of disciplines, including history, education and sociology; and examines themes as varied as Australian immigration, workplace change and literacy learning.

Laurie Gerber has a background in Asian languages, and was a central figure in SYSTRAN Software's Chinese-English and Japanese-English machine translation development efforts from 1986 to 1998. In addition to lexicography and translation module development, she helped to introduce innovations and modernisations in analysis module design and linguistic programming style, and was appointed Director of R&D in 1995. Also in contact with users, Ms. Gerber developed an interest in usability issues for language technology. In 1999, Laurie left SYSTRAN to enter a Master of Computational Linguistics program at the University of Southern California, completing the degree in May 2001. While at USC, she worked as a research assistant at the university's Information Sciences Institute (ISI) on such diverse projects as Chinese parser development, the ISI question answering system, 'WebClopedia', and automatic multi-document summarization. She now works as an independent consultant on language technology implementation, and business development for commercially viable prototype language technologies.

Gus Gollings is a project manager for the C-2-C System at Common Ground Publishing. His work covers a broad range of disciplines, from software interface engineering to research and writing to pedagogy. Recently his focus has been directed toward establishing the metadata regimes and standards required for the universal manipulation and auditing of digital texts.

Michael Singh is a Professor of Language and Culture at RMIT University and the Head of Department of Language and International Studies. He has responsibility for the development of the Globalisation and Diversity Research Concentration. His research in the areas of multicultural education, Asia literacy and Indigenous curriculum issues has involved exploring a range of socially critical practices for teachers: strategic re-interpretation, counter-construction, standpoint pedagogy, mimicry, critical literacy and contrapuntal reading. He is currently finalising a book-length manuscript entitled *Appropriate English: The Business of Teaching Global/Local Englishes.*

Atsushi Takagi has been teaching Japanese language and culture in Australia since 1983. He has developed a strong research interest in computer-mediated communication in the cross-cultural pragmatics context, and the use of multimedia for teaching Japanese. He is currently a PhD candidate in Applied Linguistics at University of Melbourne.

Dr Christopher Ziguras is a Research Fellow in the RMIT Globalisation and Cultural Diversity Research Concentration. His research interests are diverse, but in recent years have focused on the regulation of transnational education, new information and communication technologies in international education, the internationalisation of higher education, and effects of mediated self-care advice on health consciousness. Before joining RMIT he was a research fellow at the Monash Centre for Research in International Education. In 1998 he completed a doctoral thesis on the cultural politics of self-care promotion after working for several years as a freelance writer of educational and corporate video and multimedia.

Chapter 1

GLOBALISATION, MULTILINGUALISM AND THE NEW TEXT TECHNOLOGIES

Bill Cope

The word 'globalisation' describes many of the contradictory forces driving change at the beginning of the twenty-first century. Some of these forces are socially and economically objective; others are highly emotive and political. The more objective forces of globalisation are palpable, and their rapid rise can be measured by:

- the increasing proportion of total production which daily and yearly crosses borders;
- the proportion of local capital which is not locally owned;
- the increasing amount of human movement in the form of international travel;
- the sheer scale of labour and refugee migrations which have created multicultural societies around the world;
- the rapid increase in direct, person-to-person international communication in the form of mail, telephone calls and electronic communications; and
- the movement of cultural products and ideas across borders through the media and publishing industries.

Globalisation also manifests itself as a highly emotive force, and in this form it has become part of an undeniably political reality. The emotive force of globalisation is no less real, even if it is not so clearly measurable. We can count how many international phone calls are made in a day, or the number of books imported and exported. But it is not so easy to measure people's level of wellbeing or anxiety around the ideas and realities of globalisation, even though the political effects of the anxieties are very real.

In technological and economic terms, some people see globalisation as a harbinger of progress: as a source of investment; as a carrier of new technologies which improve people's work and home lives; and as a force which gradually improves the lot of the poor even though it seems to improve the lot of the rich more rapidly. However, others see these technological and economic forces as regressive – as the root cause of foreign domination and the diminishing

1

of local self-governance and democracy; as a displacer of local skills and technologies; and as a force that only serves to exacerbate disparities of wealth and privilege.

In cultural terms, the argument is no less heated. Some believe the cultural forces of globalisation help create open, cosmopolitan, multicultural societies, and a sense of citizenship built on a world-consciousness capable of handling issues which cannot or will not be tackled adequately by nation-states – environmental sustainability, human rights and fair trade. However, others believe that the cultural forces of globalisation ride roughshod over local cultures and languages, even to the point of destroying them.

The publishing and communications industries find themselves at the heart of the processes of globalisation, torn between alternative interpretations of its meaning and significance. This is an area where we may well ask the question of whether global technology drivers enable, or whether they make things more difficult for local industries. In a world economy and a competitive local economy, there is often little alternative but to invest in the best and latest tools, and in a small country like Australia these are almost entirely imported. But this very act of importing technology has immediate consequences. It means that an importing country has no particular technological edge over the source countries and other countries who have imported the same technologies, whilst at the same time being disadvantaged in relation to low labour cost countries, whether they choose to invest in the latest technologies or not. Also, the technologies are often highly loaded in cultural terms, and one example of this discussed in this book is the dominance of roman scripts and the English language through the ASCII (American Standard Code for Information Interchange) character framework, which is still the basis of most computing and text manufacturing technologies. Technology, all too easily, becomes an ally of the cultural drivers of globalisation that often seem to do enormous damage to local cultural and linguistic diversity. This technology also seems to work in concert with commercial drivers. Australian publishing and printing is increasingly dominated by global companies whose interest is bottom lines more than it is local creators, and whose investment decisions are often made in the capitals of the Anglophone publishing world: New York and London.

This book sets out to argue two things: first, that the technological, commercial and cultural forces of globalisation are moving into a very complex phase in which the effects on the Australian publishing industry may be in some senses counter-intuitive, and certainly not those predicted in many of the more emotive and political responses of recent times. The forces of globalisation need not fortify and extend the technological, commercial and cultural domination of multinational corporations, the United States and the English language. Or, at the very least, they might not *only* do this. Equally, they could be agents that foster increasing cultural diversity, greater local commercial autonomy, and the revival of local and ancestral languages and cultures. In this book, we illustrate this latter possibility with one small example – the evolution of multilingual text creation technologies during the most recent phase of globalisation.

The second aim of this book is to discuss the technological, human skills and enterprise possibilities for Australia – a small, multilingual country in an economically dynamic region where the English language is becoming less important. If America maintains its dominance in text manufacturing technologies centred around the English language, which surely it will, we should focus our energies on technological solutions for a multilingual world. And rather than being poor technological cousins in an English speaking market already saturated with print and text, we should position ourselves as savvy technology providers in non-English and non-roman script markets where the demand for print and text is, compared to the English speaking world, as yet barely developed.

GLOBALISATION AND LANGUAGE

English, the supporters and opponents of globalisation argue with equal vehemence, is becoming a world language, a *lingua mundi*, as well as a common language, a *lingua franca*, of global communications and commerce. An estimated one billion of the world's population now speaks English, the majority as a second language. Meanwhile, the world's language diversity is dramatically decreasing. It is estimated that one and a half Indigenous Australian languages disappear every year, as the last speakers of those languages die. As Michael Singh points out in Chapter 9 of this book, at the current rate, between 60 and 90 per cent of the world's 6000 languages will disappear by the end of this century.

Yet here is a paradox: recent developments in information and communications technologies may reverse this trend. Just as English appears to be becoming more important, there are signs that it may become less important. The proportion of the world's websites in English has been in permanent decline since the invention of the Internet as an almost English-only place. The proportion of the books published in English is reducing as demand for books slowly grows in rough proportion to population growth in English speaking markets, whilst at the same time massively unmet needs are gradually satisfied in rapidly growing and modernising non-English language markets, such as those of Asia.

More importantly, however, new technologies make language difference a less important factor in social communication and multilingualism the norm. We will illustrate this with one simple example: ecommerce enabled banking using a variety of machine interfaces – the automatic teller machine (ATM), Internet banking or automated telephone banking using computer-generated audio. Banking is a complex modern form of social communication. I have money, I give it to the bank to keep, and when I ask for some of it back, I make a formal application, and, if approved, the bank says so and gives it to me. This is a kind of conversational structure, carried out in the traditional shopfront bank by a complex array of written documentation, supported by actual conversations which frame the details of the transaction and give context to the written documentation. In the world of banking before electronic commerce, this was a heavily language bound activity. You had to fill out a withdrawal slip that was almost invariably only available in the 'national' language of the bank, and then speak to a teller in that language. Occasionally, in deference to multiculturalism and to make a mark on niche markets, banks would make sure there were some bilingual tellers, in order to conduct transactions with Japanese tourists, or to serve immigrant languages heavily represented in a local neighbourhood. But there were practical limits to this, the principal of which is the number of languages that can be practically serviced by a local bank.

Ecommerce enabled banking – the ATM, online banking, and automated phone banking – changes all of that. Various highly routine and predictable conversations, such as the 'I want some of my money' conversation, or the 'I want to know how much money I have left' conversation, do not really (despite appearances) happen in English. They happen through a translation of the routine operation of withdrawing funds, or seeking an account balance, into a series of computer-generated prompts. The way these prompts are realised in a particular language is arbitrary. There is nothing particular about the language of the conversation. Semantics and grammar, or meaning and information structure, are everything. The logic of the communicative exchange now operates above the level of language. Various 'banking conversations' are constructed as a universal, transnational, translinguistic code (actually, computer code, because the customer is talking to the bank's computer), in which the language manifestation of that code is, in a communicative sense, trivial. You can choose any language you like at the beginning of the online banking session and the visible 'tags' describing the effect of pressing alternative buttons will be translated into your language of choice. There is nothing to stop this being in any script; or the screen swapping its directionality if you were to choose Arabic; or non-written and non-visual interfaces, such as Braille, or interfaces translating audio to text. The ATM and telephone 'phone banking' do the same thing, working off the same ecommerce abstracted text. The rendering of the meaning and even the words of the text can be different; but the meaning-structure and semantics are the same. The business of making the banking service available in another language is as simple as translating the tags – a few hundred words at the most, and putting them into the system as writing or recording computer-generated speech for telephone banking. Once, the grammar of language was the entry point into the grammar of banking – if the customer and the bank were not able to operate competently in the same conversational, written and thus cultural world, there could be no transaction. Banking was a language-specific game, no less and no more, and the prescribed language or languages were a non-negotiable precondition for playing the banking game. However, in the world of ecommerce, the grammar of banking is created first, and this grammar can be realised in any language. Not only is this a completely different way of doing business; it is also a totally new way of thinking about communication and, even more importantly, making communication work.

Coming back to the big-picture questions of globalisation, this example captures something quite contradictory. On the one hand, billions of people have been drawn into the culture of ATMs since they were introduced in the last quarter of the twentieth century. To use a term defined and developed by the linguist Jim Gee, they have become proficient speakers of a 'social language' (Gee 1996), which we might call 'global ATM' or 'electronic banking'. The particular language-form in which this social discourse is realised for a particular transaction is, measured in terms of human action and social meaning, an arbitrary and trivial accident of birth. Yes, the culture of electronic commerce and modern banking is taking over the world, making the world the same, and doing it multilingually might be seen as a kind of ploy. But this move does meet one of the accusations of the forces opposing globalisation – it improves access for outsiders. And, at least for this kind of communication, it has become less necessary to learn the dominant language. Now you can play the global banking game, but you don't have to homogenise to be in it. You can be in a country where your language is not spoken in banks, and it doesn't matter because you can go to an ATM or ring telephone banking and deal with computer-audio, or, if needs be, a live operator in a call centre somewhere in the world who speaks your language. This is just one small and symptomatic example of the way in which new communications technologies will support language diversity, and make it less important in many settings to know a *lingua franca* such as English.

SECOND-GENERATION DIGITAL TECHNOLOGIES

In Chapter 2 of this book, Michael Singh introduces the notion of second-generation digital technologies. He argues that these will be pivotal in the multilingual knowledge economy.

The first generation of digital technologies emerged in a close fit with monolingual publishing. They were based on ASCII rendering of the Roman alphabet. If translation were to occur, it would be through another publisher who purchases rights for a different market and who republishes the work there. And the rendering technology – the printed book with its heavy upfront costs and economies of scale – favoured large languages and affluent markets.

Second-generation digital technologies change much of this dynamic in five areas:
1. The development of new font rendering systems.
2. The convergence of linguistic and visual text creation tools.
3. Text discovery and text structuring systems based on multilingual metadata.
4. Machine translation and machine-mediated human translation.
5. Flat economies of scale in digital text rendering.

1. NEW FONT RENDERING SYSTEMS

The ASCII framework consists of 94 characters – upper and lower case in the English alphabet, numerals and punctuation marks. An 8-bit character encoding system is capable of storing each of these letters as a unique pattern of up to eight zeros and ones. A whole character in a simple character set like those used for English normally 'costs' the space in memory of 1 byte, the basic unit for measuring memory and file sizes on a computer. These characters can be rendered to screen or to print as a series of dots (pixels), the number of dots depending on the image clarity required. An 8 bit (1 byte) encoding system, however, cannot represent more than a theoretical 256 characters. To represent languages with larger character sets, such as those of languages whose writing system is ideographic, specialised 2 byte systems were created. However, these remained, for all intents and purposes, separate, and designed for localised country and language use. Extensions to the ASCII 1 byte framework were created to include characters and diacritica from languages other than English whose base character set was Roman. But non-Roman scripting systems remained in their own 2 byte world. As the relationship between each character and the pattern of zeros and ones is arbitrary, and as the various systems were not necessarily created to talk to each other, different computer systems were often incompatible with each other.

For a second generation of digital technologies, a universal character system has been created called 'Unicode' in which every character and symbol in every human language is represented in a consolidated 2 byte system (www.unicode.org). The 8 zeros and ones which represent the 26 lower case letters of the Roman alphabet are now embedded in a new sixteen-bit character encoding, and are now a mere 26 characters amongst the 94,140 characters of Unicode 3.1. These Unicode characters not only capture every character in every human language, they also capture archaic languages – Linear B, a precursor to Greek found on clay tablets in Crete, has recently been added. It also captures panoply mathematical and scientific symbols. It captures geometric shapes often used in typesetting (squares, circles, dots and the like), and it captures pictographs, ranging from arrows, to international symbols such as the recycling symbol, to something so seemingly obscure as the 15 Japanese dentistry symbols.

The potential with Unicode is for every computer and every printer in the world to render text in any and every language and symbol system, and perhaps most significantly for a multilingual world, to render different scripts and symbologies on the same screen or the same page. The shift from a 1 byte to a 2 byte system increases the computer's memory requirements and decreases its speed proportionally, but measured against the pace of hardware developments in these areas, this shift is not important.

2. THE VISUAL AND THE TEXTUAL

The elemental modular unit for representing written language in the Gutenberg system was the character, and until the digitisation of text at the end of the twentieth century, this remained the elemental unit. Types were cast as separate characters and then assembled into words and lines on formes. And the elementary modular unit for representing visuals was the whole image – originally a hand engraving, and most recently, a photoengraving. These technologies and processes for manufacturing printed words and images were entirely different. In fact, as argued in Book 1, until offset printing, it made a lot of practical sense to keep words and images separate. In the case of a book, for instance, this was achieved by printing the text and the plates in separate sections.

Digital technologies make it remarkably easy to put the images and words together, and this is in part because text and images are built on the same elemental modular unit. The elemental unit of computer-rendered text is an abstraction in computer code made up of perhaps eight (e.g. ASCII) or sixteen (e.g. Unicode) bits. This is then rendered visually through the mechanised arrangement of dots, or pixels (picture elements) – a smallish number of dots rendering the particular design of the letter 'A' in 12 point Helvetica to a screen, and many more dots when rendering the same letter to a laser printer. Images are rendered in precisely the same way, as a series of dots in a particular combination differentiated range of halftones and colours. Whether they are text or images, the raw materials of digital design and rendering are all bits and pixels.

Thinking of the consequences of this change narrowly, in terms of the tools of the text creation trade, typesetting no longer even happens by itself. It has been replaced by desktop publishing in which textual and visual design happen on the same page for rendering on the same page. Even typing tools, such as Microsoft Word, have sophisticated methods for creating (drawing) and combining images.

More broadly, this convergence of linguistic and visual text creation tools facilitates and supports a shift in our communications environment which has been characterised by Gunther Kress as a movement towards visual representation and away from language (Kress 2001). We are living in a world that is becoming less reliant on words, or more precisely, a world in which words have to stand on their own, as though they are merely visual prompts. Sometimes the communication has become purely visual – it is possible to navigate an airport using the international pictographs. Other times, the visual and the linguistic are powerfully interwoven in a common communicative framework (Cope and Kalantzis 2000).

Moving away from language, or moving away from language alone, is one aspect of globalisation and multilingualism. Such a shift is a practical response to globalisation in the case of airport signage where it is simply impossible to operate in the language of every traveller. It also reduces language to a less important part of the communications equation. The real meaning of a technical manual is in its structure and diagrams; and if the design of the manual text is kept to a minimum, it is a relatively inexpensive task to translate labels and text and insert this into the digitised pages. The real meaning (and most of the design work) in an architectural glossy is in the images, the plans and the technical data. Introductions, captions and other text are easily translated and inserted into the source file. Communications, in other words, are built on linguistically open visual templates, in which the text is no more than a secondary component.

Ron Scollon speaks of an emerging 'visual holophrastic language'. He derives the term 'holophrastic' from research on young children's language in which an enormous load is put on a word such as 'some' which can only be interpreted by a caregiver in a context of visual, spatial and experiential association. In today's globalised world, brand logos and brand names (to what language does the word 'SONY' belong? he asks), form an internationalised visual language. A visual holophrastic symbol brings with it a whole pile of visual, spatial and experiential associations, and these are designed to cross language barriers (Scollon 1999).

3. MULTILINGUAL METADATA

Metadata schemas use 'tagging' frameworks to describe the content of documents. In the case of documents locatable on the Internet, Dublin Core is one of the principal emerging standards (http://dublincore.org), and is typical of others. It contains a number of elements: title, creator, subject, description, publisher, contributor, date, resource type, format, resource identifier, source, language, relation, coverage and rights. The schema is designed to function as a kind of 'catalogue card' for the Internet, so that it becomes possible, for instance, to search for Benjamin Disraeli as an author (creator) because you want to locate one of his novels, as opposed to writings about Benjamin Disraeli as a British Prime Minister (subject) because you have an interest in British parliamentary history. The intention of Dublin Core is to develop resource discovery tools more sophisticated than the current search tools, which can do little more than search indiscriminately for words and combinations of words.

This kind of a framework supports multilingualism in several ways. Firstly, it does not assume ('naturally') that the data will be in any particular language, presumably the language of point of search and entry. Even if you are working through an English language searching and cataloguing framework and the data happens to be in English, English will be specified. Data in any language sits within a common, multilingual resource description framework. Second, the tags are progressively being translated into a variety of languages. Although this may seem a small move in a practical sense, merely involving the translation of a few dozen terms for each additional language, conceptually it is a huge move. In fact, it turns a linguistically expressed term into a mere 'token' of a core concept that exists above and beyond any particular language. To achieve this, thesauri of terms are created in parallel across multiple languages in order to stabilise language tokens for each of the core concepts. This means that the metadata qualifier for locating an author is not really 'creator' at all, but the concept of creator which has been translated into however many languages, and is always marked by the same word from each of these languages.

The same potential exists for physical books. The online information transfer language, ONIX, is being developed to create a common platform for B-2-B and B-2-C ecommerce transactions for publishers and booksellers. However, being an initiative of the US and UK Publishing Associations, this is as yet an English-only framework. Creating one world for ecommerce in books is a matter of translating the ONIX tags.

The most advanced of contemporary dedicated book creation tools, such as DocBook, build text around the book as a structural architecture, rather than the book as a physical product whose structure is merely seen in the visual rendering. They also radically blur the distinction between data and metadata. 'Creator' is not only a metadata concept; it is possibly an instruction to render the creator's name on the title page of the book. Metadata does not only sit outside of the data; it creates rules that automate the manufacture of alternative renderings. The printed book will be rendered in one way, the html text another, the ebook for a reading tablet another, the talking-book another. But the creator will always be the creator and the rendering will always make it clear that this particular set of words does, in fact, represent the name of the creator. DocBook supports Unicode characters within the data. The revolutionary step (and technically quite straightforward) will occur when the tags that structure the text – title, chapter heading, subheading, table of contents and so on – are available in multiple languages and scripts, and the translation of these tags is stabilised by convention. Then it will be possible to write a book in any language or script in the world, and print it on any printer.

In a globalised and multilingual world, Ron Scollon argues, social languages or discourses are more similar across languages than within languages (Scollon 1999). The way academics write in English and Japanese is very similar in terms of the structure of their texts and the ways those texts describe the world. A DocBook framework for structuring academic text, which may include elaborate referencing, keyword, indexing and abstracting apparatuses, will work across languages if the tags are translated, and this is because the most important thing about the discourse, and the final document, does not sit inside a particular language. Text structured and rendered using tools such as DocBook is a perfect platform for multilingual, multi-script publishing in communities more and more defined by discourse (what they want to do in the world, as expressed in peculiar ways of communicating about the world) than by the accident of mother tongue.

4. MACHINE TRANSLATION

Into this environment, more than fortuitously, steps machine translation. Despite the enormous complexities of human language, machine translation tools can now provide the gist of the meaning of a written text. Some of these, such as AltaVista's Babelfish, offer instantaneous online translations free. The Babelfish service is powered by Systran, one of the founding companies in the machine translation field, and offers nineteen language pairs (http://babel.altavista.com). Machine translation is getting progressively better, even though it may never equal human translation.

The gist of a text might be sufficient to justify a decision to pay for human translation. Systran, for instance, offers an online human translation service in alliance with Berlitz Globalnet in which any document below 2000 words will be translated and emailed back within twenty-four hours. Machine translation also makes the job quicker, particularly when it is linked to texts which have been deliberately designed for multilingual delivery, using such methods as controlled vocabularies, based on the translation of technical terms peculiar to a particular field.

The range of language pairs available in online human translation services is still limited. The potential, though is enormous. A researcher working on frogs in a rainforest in Laos wants to make a comparison with a similar tree frog in a Brazil. Using a resource discovery framework which operates multilingually, this researcher locates a chapter of a book about the Brazilian frog in Portuguese, decides from a machine translation of gist that this is an important reference, then finds a Portuguese-Laotian translator amongst the thousands of registered translators representing every imaginable language pair. The publisher then captures this translation forever.

Highly structured text frameworks for capturing data and metadata, along the lines of DocBook, will also help in the machine translation process, as well as, even more simply, making text of printed books available online in a format in which it is possible to undertake machine translation faithful to the text's structure and form.

5. ECONOMIES OF SCALE IN DIGITAL TEXT RENDERING

Finally, the flat economies of scale in the new world of digital print and etext convergence discussed in Book 2.1, make it possible for small languages and small cultures to express themselves through the book form. With the kinds of mass customisation systems envisaged in Book 2.3, the per unit cost of a digitally printed book is constant regardless of the run length. There is no effective distinction between long run and short run, niche markets and mass markets. If there is only a small market for a book of Chinese diasporic poetry, the cost of manufacture and publishing will no longer be an inhibitor to that book's production. The text creation process will occur online in an environment in which all the tagging and mark-up required to render the book to print and other formats can be seen in Chinese, in which the text itself can be written using Unicode, and in which the book, once published will be available in any online bookstore in the world.

To take another example of how this may provide opportunities and access for small cultures and small languages, East Timor faces a daunting challenge to replace Indonesian as the language of instruction in schools. The cost of producing textbooks is prohibitive. But a teacher with a single computer with metadata maker and structured text framework translated in to Tetum, one of the main local languages, will be able to produce precisely the number of books needed, printed cheaply using digital print technology. It is these kinds of technologies, these kinds of possibilities, which might breathe new life into small and declining languages.

MULTILINGUAL PUBLISHING IN A GLOBAL KNOWLEDGE ECONOMY

This book explores multilingual publishing opportunities in the light of changing text-manufacture technologies.

In Chapter 2, Michael Singh introduces the notion of second-generation digital technologies and discusses their potential role in the global knowledge economy. The opportunities are enormous – amongst the 189 million speakers of Bengali or 63 million speakers of Tamil, for instance, whose capacity to purchase books and demand for books is bound to grow more rapidly than the mature markets of the Anglophone world.

13

Christopher Ziguras and Melissa Brown analyse current state of Australian multilingual book publishing in Chapters 3 and 4, including a number of case studies of successful multilingual publishers, printers and online bookstores. Here they highlight the possibility that we are not exploiting the enormous demographic possibilities for Australia in a multilingual world. According to the 1996 census, 2.5 million people speak a language other than English at home. Yet most book exports are of English text. Although there is some licensing of translation rights to books created in Australia, and some Australian publishers have set up international alliances or even offshore branch offices, there are still enormous opportunities that Australian publishers and printers have barely noticed, let alone utilised.

In Chapters 5 and 6, Gus Gollings explains some of the technological developments in the areas of multilingual typesetting and character encoding. He begins with a history of the development of typesetting, and moves on to discuss the evolution of computerised text coding. This provides the foundation for an explanation of the technical basis and potential significance of the move from ASCII to the Unicode standard as the basis for all digital character encoding. Atsushi Takagi extends this discussion in Chapter 7, examining a number of practical typesetting issues in non-roman scripts.

Chapter 8, by Laurie Gerber who has worked for Systran, discusses the future of translation in a digital environment. She analyses the technical issues of machine translation, including language modelling techniques being developed in computer science and computational linguistics will she believes will make machine translation a viable option for many more types of translation in the future. She also discusses the inherent limitations of machine translation, pointing to an increasingly globalised future in which machine assisted human translation will be of increasing significance.

Michael Singh concludes the book with a reflection in Chapter 9 on the skills development requirements of a country like Australia if it is to avail itself of the opportunities available in a global, multilingual knowledge economy. Whereas the publishing and printing industries consolidated historical processes which attempted to force single official languages on nation-states, and which assisted the rise of English to global domination, second-generation digital technologies may well turn the tide in favour of language diversity, and thus cultural diversity.

14

And herein lies the paradox of globalisation, at least in its most recent incarnation. The universal processes of globalisation bring the peoples, cultures and languages of the world closer together. But bringing them closer together does not have to make them more homogenous. Indeed, the emerging technical tools of digital text creation and manufacture enable quite the opposite – the revival of small cultures and languages that languished in the earlier phase of globalisation with its first generation digital technologies. In the new world of text creation, all human symbologies work within a single technical frame of reference; the visual increases in relative significance relative to the linguistic; metadata schemas for finding and structuring text operate above the level of language and thus across multiple languages; machine translation works reasonably well while online human translation works extremely well; and niche markets become just as viable as mass markets as a result of technologies and business practices which facilitate mass customisation. This is not a globalisation of cultural and linguistic homogenisation Rather, to use Michael Singh's term, the dynamic becomes one of globalised localisation or localised globalisation.

REFERENCES

Cope, Bill, and Mary Kalantzis, eds. 2000. *Multiliteracies: Literacy Learning and the Design of Social Futures.* London: Routledge.

Gee, James Paul. 1996. *Social Linguistics and Literacies: Ideology in Discourses.* 2nd Edition ed. London: Taylor and Francis.

Kress, Gunther. 2001. 'Issues for a Working Agenda in Literacy'. In *Transformations in Language and Learning: Perspectives on Multiliteracies,* edited by M. Kalantzis and B. Cope. Melbourne: theLearner.com/Common Ground Publishing.

Scollon, Ron. 1999. Multilingualism and Intellectual Property: *Visual Holophrastic Discourse and the Commodity/Sign.* Paper presented at GURT 1999.

Chapter 2

GLOBAL BOOK PRODUCTION FOR THE MULTILINGUAL KNOWLEDGE ECONOMY

Michael Singh

English was supposed to be the global language of the digital information age. Why then are we seeing the rapid emergence of a second generation of digital technology that enables publication and printing in a diverse range of languages and scripts? The triggers for the emergence of this new technology are multiple, complex and even contradictory. This second generation of digital technology exists due to a clear market demand for products and services in many languages other than English. This market demand is emerging because English-only products and services are inadequate for accessing the diverse range of knowledges and niche markets now available globally. This new generation of digital technology is emerging in response to resistance to the project of globalising English.

Although this resistance movement is mobilised around the issue of language rights, it is not unrelated to the complex and contradictory resistance to the imposition of other aspects of globalisation from above. This new technology is gaining support among consumers and investors who have an ethical interest in sustaining the world's linguistic diversity and associated knowledge resources.

Book publishing and printing are important agents in explaining and creating the world's emerging multilingual knowledge economy. The work of authors, publishers and printers, educators, librarians and booksellers is part of the trade in knowledge distribution surrounding globalisation, the sustainability of bio-linguistic diversity and technological innovations. At the beginning of the 21st century, the second generation of digital technologies now makes it possible to write and print books produced by and for many of humanity's diverse language markets, including Armenian, Cherokee, Lao and Tibetan. Given that the middle class component of speakers of languages such as Bengali (189,000,000), Tamil (63,075,000) or Telugu (66,350,000), is as large if not larger than that of the comparable market in Australia, what barriers have to be addressed to access markets such as these?

This chapter provides a preliminary outline of issues relating to the production of books for a global *multilingual knowledge economy* and how these are now being addressed using *second-generation digital technologies*. Specifically, it addresses the following questions:

- What is the multilingual knowledge economy?
- What is the rationale for an open multilingual platform for publishing?
- What are the political, ethical and cultural risks related to the management of the globalisation of English?
- What is the projected status of English in book publishing?
- What is the worldwide trend for printing in English and other languages?
- What are key cultural issues in translation?
- What are the advantages of the Australian publishing and printing industry taking on the issue of the sustainability of the world's linguistic diversity?
- What is the possible role for the Australian industry in multilingual publishing and printing?

Before addressing these questions, the following vignette provides some insights into what is already possible using second-generation digital technologies to situate the Australian publishing and printing industry within the emerging multilingual knowledge economy.

A VIGENETTE: BOOK PUBLISHING FOR THE MULTILINGUAL KNOWLEDGE ECONOMY

Ali Darwish's (2001) guide for Arabic translators and interpreters of English was produced by Writescope Pty Ltd. Up until the publication of *The Translator's Guide*, there was not a book which explained the dynamics of translation or provided a clear framework for translation between Arabic and English. Ali Darwish's book makes a useful contribution to the Arabic library internationally and domestically by redressing the lack of serious works on translation theory and practice. Moreover, just as *The Translator's Guide* addresses the ethical obligations required of professional interpreters and translators, there is a growing interest in ethical investment as a risk management strategy across a wide range of businesses and industries.

Further, *The Translator's Guide* alerts us to the close relationship between globalisation and the creation of a global multilingual knowledge economy via the book publishing and printing industry and the significance of transnational language communities

Writescope is a small business of four consultants specialising in technical communication, information design, translation and communication. *The Translator's Guide* (2001) was one of the publications the company undertook in collaboration with Algethour. Ali Darwish wrote and formatted the book to the camera-ready stage on an IBM compatible PC using the Arabic versions of Microsoft Word 6.0 and later Microsoft Word 2000. These days Microsoft Word is more than adequate for Arabic and most Latin-based scripts. Using Microsoft Word instead of Adobe Frame Maker for example, reduces the cost of initial investment in powerful packages. With the use of customised templates and macros, Microsoft Word does the job. This process was aided by the author's expertise as a technical communication and information development specialist.

The development process was iterative and followed a structured methodology developed by the author and adopted by Writescope. In the final stage of development, the book was reviewed and revised several times by the author and two other consultants. The book was typeset on an IBM compatible PC using Microsoft Word. The final master was then burned onto a CD in PDF format (Adobe Acrobat 5.0). The CD was handed over to Algethour's Ali Abusalem who coordinated and subcontracted the printing process to a digital printing company. Digital printing technology was used so as to enable printing on-demand.

The local market for this book is very small. This is why Writescope is seeking to work with overseas publishers (in the Arabic speaking countries, especially Lebanon and in Germany) to publish the book jointly. Through networking and an Internet presence, Writescope has been able to establish links with major publishers and distributors overseas. Writescope is now in the process of streamlining its processes with Algethour and other providers, and is also exploring further possibilities for multi-lingual publishing (for example English–Arabic, Arabic–Spanish, and Arabic–French).

At this point it must be recognised that Australian multiculturalism has been successful in encouraging bilingual Australians to maintain and develop their transnational links. For instance, the Algethour Cultural Association, which is facilitating the Arabic-speaking community to find its place in Australian multicultural society, does so by assisting with the printing and publishing of books in Arabic both here and abroad. The Association was established in 1999 to publish a cultural journal, *Algethour*, which focuses on inter-ethnic issues, and encourages budding writers from different ethnic backgrounds to contribute poetry and short stories. To enhance the journal's quality assurance the Association has established a review panel to evaluate manuscripts. It is now involved in subcontracting this work to local printing houses, such as Writescope, liaising with the printers, and handling the processes of proof reading, collation and assembling of the electronic and hard copies of the book.

In addition to the organization of printing, the Algethour Cultural Association also makes sure the book meets publishing standards and assists with book promotion and distribution. If such enterprises prove successful and expand there may be a competitive advantage for Australian printers such as Writescope to provide small print runs for books on demand.

The production of *The Translator's Guide* provides a tangible illustration of the possibilities that are achievable now in the Australian publishing and printing industry. By making productive use of the resources of corporate multiculturalism and second-generation digital technologies, Writescope has been able to explore new market opportunities in the multilingual knowledge economy. Customers can now easily order Ali Darwish's guide for Arabic translators and interpreters from net-based booksellers.[1]

WHAT IS THE MULTILINGUAL KNOWLEDGE ECONOMY?

As a result of the transformations created by second-generation digital technology, Australian printing and publishing is now confronted by the challenge of the emerging multilingual knowledge economy. In this dimension of the globalised economy, it is the knowledge contained within the world's many different languages that is now a site of conflict over intellectual property rights and

---

[1] To obtain one or more copies all you need to do is email your request to the Publicity Manager at P.O. Box 418 Patterson Lakes, Victoria 3197, or at writescope@surf.net.au.

wealth creation. As knowledge and information become central to economic life globally, these high value-added goods and services are increasingly tailored to the niche markets created by cultural and linguistic diversity. While once there may have been a competitive advantage in the concentration of knowledge and intellectual resources in the English language, second-generation digital technology now makes it much easier for many different languages to enter the global marketplace and establish distributed and networked markets.

Moreover, the emergence of a global multilingual knowledge economy, which gives priority to information processing services over the production of material goods, has implications for the nation's meta-economic functions such as education and training policy. Accordingly, investment in the education and training of editors, proof-readers, translators, web managers and cross-cultural sales representatives is seen as integral to the survival of the Australian publishing and printing industry in this multilingual knowledge economy.

Language as a repository for knowledge of technology, art, music and much more, has driven human endeavour across the generations. Much has been learnt, and still must be learnt, from the Inuit people about the Arctic climate, as well as the management of marine resources from Pacific Islanders. The value and breadth of knowledge contained in the many languages of the world is still unknown. 'The next great steps in scientific development may lie locked up in some obscure language in a distant rain forest' (Nettle and Romaine, 2000, p. 16). The acquisition, accumulation, maintenance and transmission of the resources contained in the world's 6,000 languages plays a crucial role in the global knowledge economy. The problem of language extinction raises key concerns about the death of knowledge, and that is of great importance to the emerging knowledge economy. For instance, the extinction of a local language means the death of intimate habitational knowledge – understanding the land, water, plants, and animals specific to that environment. This would be a loss to science and industry of knowledge that could benefit from learning how to manage ecosystems more effectively.

In the knowledge economy, multilingualism is not something to be avoided or even done away with, but rather the very resource base for the economy. Nettle and Romaine (2000, pp. 56–77) provide an account of the encyclopaedic knowledge that the world's economy has already lost through the extinction of different language and the associated decimation of its speakers. In the case of Hawaiian people, the world has lost intimate and in-depth knowledge of the fish of their islands – fish behaviours, fishing practices and technology – knowledge yet to be documented by scientists. The loss of the world's linguistic diversity is affecting scientific and medical knowledge as well as social and economic progress. Based on Crystal's (2000, p. 44) account, it is reasonable to argue that a monolingual knowledge economy is at a disadvantage, depriving business of a range of niche (linguistically diverse) markets. One way of increasing the resource base of the knowledge economy is to access the knowledge resources of other languages. In the supply chain, the work of globalising business has now taken on processes of localisation whereby products and services are adapted to suit target languages and cultures.

WHAT IS THE RATIONALE FOR AN OPEN MULTILINGUAL PLATFORM FOR PUBLISHING?

The question that arises from this vignette is: How was it possible to publish a book in Arabic in Australia that is now available via on-demand printing and may be sold into markets in the Middle East and Germany? The answer lies in the development and market-driven application of open multilingual metadata platforms that allow for multilingual publishing and printing. That is to say that multilingual book production in Australia for globally segmented markets is now possible due to the creation and business application of second-generation digital technology. The use of this technology is now needed to access diverse language markets beyond the already saturated English-reading market.

Metadata schemas, based on any of the accepted or emerging standards, capture an amazing array of information about a given book, including the language in which it is written. It is these metadata platforms that have made possible the multilingual capacity of second-generation digital technology, and is doing so at an increasing rate for an increasing range of languages in response to shifting markets needs. Neither large-scale nor niche book production and distribution is possible without it. The dominant metadata standards have from inception included a field called 'Language' that describes the language(s) of the product's intellectual content. However, it is only in the second generation of digital technology that the language of the metadata themselves have been opened-up and can now be defined in almost any written human language. There are several competing standards, each specialising in a niche.

The Dublin Core Metadata Initiative (DCMI) is an open forum engaged in the development of interoperable online metadata standards that support a diverse range of business purposes and models.[2] DCMI conducted its originating workshop in Dublin, Ohio in 1995, giving it its name. DCMI promotes interoperable metadata standards and develops specialised metadata vocabularies for describing and enabling the delivery of information.

DCMI's activities include consensus-driven working groups, global workshops, conferences, standards liaison, and educational efforts to promote widespread acceptance of metadata standards and practices.[3] Originally conceived for author-generated description of Web resources, Dublin Core has attracted the attention of formal resource description communities such as publishers, museums, libraries, government agencies and commercial organizations.

DCMI is designed to foster the interchange and reuse of resources across linguistic and cultural differences, while pursuing the challenging aim of being culturally neutral.[4] The World Wide Web Consortium (W3.org) develops the specifications, guidelines, software, and tools to support Web infrastructure, including HTML, RDF, and XML. Each of these has a role in the syntactic encoding of Dublin Core metadata. The Dublin Core benefits from active participation and promotion in some 20 countries in North America, Europe, Australia and Asia.[5]

A major problem with the earlier generation of digital technologies was the inability of their programs and/or platforms to deal with non-Roman scripts. First generation digital technologies were unable to deal with Cyrillic characters, accented Latin characters, Greek characters, let alone languages such as Egyptian or Thai that use scripts that differ markedly from the Latin alphabet. One of the more significant developments in second-generation digital technol-

---

[2] http://dublincore.org//about/index/shtml.rdf
[3] http://dublincore.org/
[4] http://ariadne.unil.ch/metadata/agreement.content.html
[5] http://www.ukoln.ac.uk/metadata/resources/dc/

ogy is Unicode, which addresses this problem, thereby opening up new market opportunities. Moving beyond the 128 character set established by the American Code for Information Interchange, Unicode has 49,194 characters covering the scripts of many world languages (and markets), including Chinese, Japanese and Korean. Unicode provides a unique number for every character in each of the major scripts of the world's written languages, irrespective of the computing platform, the software program or the language used.[6] Given the increase in the number of HTML codes, fonts and editors Unicode is able to support, it is now possible to produce a range of standardised multilingual publications as well as Web pages. Developed by the Unicode Consortium, a non-profit organization representing a broad spectrum of the computer and information processing industry, Unicode is now the officially sanctioned way of implementing ISO/IEC 10646.

WHAT ARE THE POLITICAL, ETHICAL AND CULTURAL RISKS TO BE MANAGED IN DEALING WITH THE GLOBALISATION OF ENGLISH?

The idea that the English language should achieve the status and power attached to being the world's 'global language' is not universally seen as having self-evident benefits. Let's not forget, English gained its present global dominance 'by replacing the languages of Indigenous groups such as Native Americans, the Celts, and the Australian Aborigines, and now many more' (Nettle, and Romaine, 2000, p. 15). The acclaimed author Amitav Ghosh is ambivalent about writing and publishing in English. Ghosh duly recognises both the influences and resources available in a language he now regards with suspicion as well as the need to celebrate the literature of other Indian languages. It was this ambivalence that led him to withdraw his anti-imperialist novel from the Commonwealth Writers' Prize in March 2001, stating:

> So far as I can determine, *The Glass Palace* is eligible for the Commonwealth Prize partly because it was written in English and partly because I happen to belong to a region that was once conquered and ruled by Imperial Britain. Of the many reasons why a book's merits may be re-

---

[6] http://www.unicode.org

cognised these seem to me to be the least persuasive (Gosh in Griffin, 2001, p. 5).

In terms of risk management, it is important for the Australian publishing and printing industry to be aware of the political and cultural resistances to the project of globalising English. Warschauer (2000, p. 155) reports that, 'throughout the world, from France to Hong Kong to Malaysia to Kenya, movements have arisen to defend national [and Indigenous] languages against the encroachment of global English.' In part, this resistance has arisen because the earlier generation of English-only digital technologies created obstacles and complications for people wanting to use their own vernacular (Dragona and Handa, 2001, pp. 52–73). Further resistance continues as speakers of languages other than English learn about the continuing racist and sexist constructions of themselves in English-only electronic and digital publications (Ow, 2000).

Language is a powerful medium and rallying point for mobilising community opposition to the cultural and linguistic homogenisation created by the globalisation of English. This oppositional project has been closely identified with the information and communications sector, including the printing and publishing industry. The global language rights movement is developing a holistic response that understands how pressure to assimilate into the world-dominating language and culture is threatening to the lives and knowledge of smaller language groups.

In Korea, for instance, resistance to English was reinforced when Microsoft attempted to buy out a local software company and thwart the development and production of a local, popular word processing program (Auh, 1999). Microsoft hoped the deal would secure a monopoly for its version of the Korean language software program. Frustrated and angry, Korean computer users and non-users rallied in a massive nationwide campaign against this instance of 'predatory globalisation' (Falk, 1999). It is important that the Australian printing and publishing industry is cognisant of the political, ethical and cultural risks associated with the globalisation of English if these are to be managed prudently.

WHAT IS THE PROJECTED STATUS OF ENGLISH IN BOOK PUBLISHING?

Second-generation digital technologies have created possibilities for multilingual publishing and printing based on a marketing commitment to access the multilingual knowledge economy, thereby helping to sustain the world's linguistic diversity and its knowledge resources. However, it is worth recalling the reasons why the first generation of digital technologies was dominated by the English language. Firstly, most of the initial users of the technologies were North American speakers of standardised US/American English. They took this opportunity to privilege their language over all others through the creation of the Internet host, computers and Homepages. Secondly, those who designed and developed the new digital technologies did so using the American Standard Code for Information Interchange (ASCII) 'which made computing in other alphabets or character sets inconvenient or impossible' (Warschauer, 2000, p. 156). Once combined, these factors increased the need for people worldwide to communicate in a common language, thereby reinforcing the dominance of English in cyberspace. To this we should add those whose political commitment is for an English-only world, or at least a world wherein English is the common language, as well as those concerned about the global marketing of English language products and services (Macedo, 2000). The latter interests once saw this first generation of digital technologies as the flagship for delivering their project of globalising English.

Not only has the English language dominated international trade, but 'books in the English language have dominated the publishing business; there are few countries in the world where English books cannot find a market of some kind' (Nettle and Romaine, 2000, p. 18). For Australian publishers and printers, this suggests potential market opportunities among the three quarters of the world's population who do not speak English as their first language – or perhaps do not speak English at all.

To date, the Internet has allowed English-speaking users of the Latin alphabet to publish their own work in their own tongue, but the second generation of digital technology will support the use of many more languages for local and trans-global communications. The creators of any new publishing and printing enterprises, or those involved in reinventing or reworking existing ones, now have to decide in which languages in the global multilingual marketplace they will produce books. They have to decide whether the consumers of books will be able to choose books in their preferred language from booksellers. Here, then, the Australian publishing and printing industry has to decide whether it can utilise the capacity of second-generation digital technology to participate and prosper in the global multilingual knowledge economy, or whether it can survive in a market restricted to English-only book production.

WHAT IS THE WORLDWIDE TREND FOR PRINTING IN ENGLISH AND OTHER LANGUAGES?

An increasing number of books are being published in East Asia and Western Europe relative to those being published in the USA and Canada. This means that there is a corresponding increase in the market share of books being published in the languages of East Asia and Western Europe within the global publishing and printing industry (Reed-Scott, 1999). China is the world's most prolific publisher and its entry into the World Trade Organization creates significant opportunities for multilingual publishing and printing in the Australia industry. Japan and Korea are important publishing centres, indicating significant potential markets. Book production in East Asia mirrors cultural, socio-political and economic developments of these countries.

China's entry into the WTO, along with its hosting of the 2008 Olympic Games and drive for widespread provision of higher education (and therefore increasing numbers of sophisticated readers) are especially significant for the Australian industry. Moreover, the increasing levels of publishing throughout East Asia reflects the increasing availability of second-generation digital technologies, and the commitment of publishers and printers to develop these markets.

Switzerland, Italy, France, Germany, the UK and the Netherlands combined, produce more books than the USA (Reed-Scott, 1999). 44% of the world's book production occurs within Europe, in a diverse range of European languages. The Eastern and Central European publishing industry and market continue to be unstable given the political and economic changes occurring there since the late 1980s. Brazil and Mexico are leaders in the publishing and printing industry in Latin America. There has also been an expansion of the printing and publishing industry throughout the region ranging from Morocco to Afghanistan, and from Turkey to the Sudan. These trends point to both the saturation of the English-reading market for books and recognition of the need to access a range of other multilingual markets.

WHAT ARE THE KEY CULTURAL ISSUES IN TRANSLATION?

In today's climate of globalisation, translation has gained considerable prominence in international communication. In multicultural Australia, translation is more vital now than ever before in providing services to, and doing business with, an ever-increasing range of speakers of languages other than English – migrants, refugees, temporary residents, tourists, businesspeople and international students. Much book content has to be adapted to local cultural, political, climatic or educational requirements. For instance, because of differences in spelling and measurement systems, US schools books are inappropriate for the Australian school textbook market. Likewise, markets for leisure books on sport and fishing differ by geographic region and the socio-economic status of the target demographic groups. Given that local requirements of given markets influence book sales in English, similar differences in other languages could create opportunities for carving out distinctive and profitable niche markets.

The Internet provides many human language translation sites including online dictionaries and e-mail translation services. Given the global nature of the Internet, electronic translation businesses offer a growing field of work. But what is the possibility of a translation system that automatically translates a text in one language into that of another language? Work is currently being done on developing computer software that address this process, as well as several other related aspects: namely machine translation, voice-recognition systems and speech synthesizers (Wallraff, 2000). AltaVista's Babel Fish provides a rudimentary free online service that hacks out rough 'text-to-text' translations for English into a 'destination language' and vice versa. Professional translators then refine these rough translations. Dragon Systems' *Naturally Speaking* provides for 'speech to speech' translation. 'Speech to text' translation systems of particular interest to authors, such as Dragon Systems' *Naturally Speaking* and Kurzwell Education Systems, use voice recognition software to turn spoken words into written words. The visually impaired can benefit from the Kurzwell 'text to speech' machine's ability to turn written words into spoken words in fifty (out of humanity's 6,000) languages.

Software to support language translation is improving, making it possible for authors, publishers and printers to produce books in a range of languages. However, the prospects for automatic machine translation of natural conversations, literary, technical or legal material is still in serious doubt. Fully accurate, high quality machine translation of any text in any language is, at this stage, an unrealisable goal. Nevertheless, it is possible to develop automated aids to assist human translators as well as to translate material provided using a restricted language code. The Systran and METEO systems have demonstrated that the translation of the restricted language materials available on the Internet is an increasing likelihood. For instance, the METEO is designed for translating copious amounts of meteorological data in bilingual Canada. A high degree of accuracy using machine-assisted translation requires the imposition of tight restrictions on the discourse domains to be translated through predictable lexicons and simplified, stripped-down syntax. Machines can be used for translating texts relating to setting appointments, making travel arrangements, recipes or manuals.

WHAT ADVANTAGE DOES THE AUSTRALIAN PUBLISHING AND PRINTING INDUSTRY GAIN FROM TAKING ON THE ISSUE OF THE SUSTAINABILITY OF THE WORLD'S LINGUISTIC DIVERSITY?

The geographical and language-based agreements, such as international copyright conventions about the right to publish works in a specific regional and language market, have been steadily diminishing as a result of the new digital technologies. For some at least, the Internet is being used to tear down the world's borders and customs checkpoints as well as reducing the inconvenience of lengthy international shipping delays. Publishers and printers are starting to feel the economic effects of efforts to maintain geographically and linguistically defined markets.

Of course, some national governments are keen for the digital technologies associated with printing and publishing to incorporate and enforce geographic market constraints and so stop digital content from easily jumping national boundaries. For instance, Nuvo-Media has produced the Rocket e-Book to support geographic limitations via a territorial rights management system. This powerful system seeks to control both marketplace segmentation as well as various forms of information control and censorship. The motives for protecting regional markets is to preserve the structuring of existing economic arrangements as well as to support nation-focused policies.

Should the Australian publishing and printing industry take on a commitment to the sustainability of the world's linguistic diversity, a competitive advantage for ethical investment in the multilingual knowledge economy may be the reward. There is a world-wide movement within business to respond to and engage with issues of corporate, social, and environmental responsibility as a necessary aspect of good business investment and risk management (G. Cazalet, gcazlet@corporate-citizenship.com.au). Instead of focusing solely on their profit margins, smart businesses are giving increasing attention to what is now called the 'triple bottom line' – the related and interdependent dimensions of their social, environmental and economic responsibilities. The idea that the world's corporate citizens can or should operate outside the complex and evolving ethical and moral frameworks expressed by communities worldwide has been contested on the streets of Genoa, Seattle, Davos and Melbourne.

Moreover, social auditing is able to expose those companies that engage in 'ethical makeovers' – changing the corporation's brand image while not substantially changing their social and environmental misbehaviour. Increasingly, investors are also raising their concerns about corporate responsibilities. Britain's Centre for Tomorrow's Company and the World Business Council for Sustainable Development are working to make social and environmental accounting an integral part of annual business reports and the performance assessment of businesses. Given the standards set for such reporting in Britain, companies in Australia have some way to go. However, the restructuring of global competition in the multilingual knowledge economy means that issues of ethical responsibility to shareholders, workers and consumers are now matters of core business.

It is within the context of mounting interest in ethical investment that the Australian printing and publishing industry might usefully consider the extinction of three quarters of the world's language and the associated knowledge death during the course of this century. The industry represents corporate citizens who could be seen as having certain responsibilities to the global multilingual knowledge economy. Given the trends in ethical investment it is likely that shareholders in the industry will become attentive to its contribution to efforts to sustain multilingualism. The obligations of corporate citizens are not limited by a contract specifying mutual obligations between themselves as employer and their employees, shareholders and customers, but because of the obligation to act in the world justly and without deception. Cheating people of their language raises questions about the obligations publishers and printers of books – which have to be in one or more languages – have towards sustaining the world's linguistic diversity. Whether these arguments are sufficient to induce the beneficiaries of today's global linguistic hierarchy – English language speakers, investors, authors, publishers, printers, editors – to support the sustainability of humanity's linguistic diversity has yet to be tested through innovations in the use of second-generation digital technologies.

Informed and active shareholders concerned about the ethics of their investments are raising questions about the social and environmental credentials of businesses associated with product such as weapons, alcohol, furs, sprays that damage the ozone layer, pharmaceuticals, mining, gambling, tourism and genetic engineering. In the USA some $2.3 trillion is managed by ethical investment funds; in the UK the figure is $A11 billion, while in Australia there is $A1 billion in socially screened investments (Kemp, 2001, p. 6). These figures are continuing to grow.

Of course, ethical investment in a diversity of socially aware companies has to be justified to shareholders by empirical evidence of corporate responsibility and performance (as measured by the Westpac-Monash Eco Index for example). Businesses such as the Australian Ethical Equities Trust, Guy Hooker's Ethical Investment Cooperative (UK) and the Ethical Investment Company (Australia) provide advice to investors on portfolios of ethical stocks. Nettle and Romaine (2000, pp. 23–24) raise a series of important questions for the Australian publishing and printing industry to consider. In the press for technological innovation and the global restructuring of its markets could there be a competitive advantage for the industry in changing its brand image and substantially changing its social and environmental behaviour by taking on a commitment to the sustainability of the world's linguistic diversity? Is consumer knowledge becoming such that many readers will refuse to buy books from publishers and printers that are not contributing to efforts to sustain the world's linguistic diversity and associated knowledge resources?

WHAT IS THE POSSIBLE ROLE FOR THE AUSTRALIAN INDUSTRY IN MULTILINGUAL PUBLISHING AND PRINTING?

The Australian printing and publishing industry is mainly positioned within the world's English reading market, although as the case studies in Chapter 7 indicate, the Australian industry does possess the capacity for multilingual publishing and printing. The creation of a 'free market' in the international book trade is concerned with opening up markets in those parts of the global multilingual knowledge economy that read languages other than English. In other words, if the restructuring of global agreements governing the book trade is a form of guerrilla marketing, then the following figures suggest that there could be a competitive advantage to be gained from engaging linguistically diverse markets. A major source of competition for the Australian publishing industry comes from the USA (with 37% of imports), while in the late 1990s UK publishers dominated 33% of the Australian market. Reflecting the offshore printing by UK and US publishers, Singapore held a 10% share of the market, Hong Kong had 8% of the total market, and China a small but increasing market segment. New Zealand is the number one export destination for Australian publishers' books. Exports to New Zealand accounted for some 35% of export sales in 1996 followed by the US (US$5 million), the UK (US$3.9 million) and Singapore (US$2.3 million). Total exports for that year amounted to US$40.9 million.[7]

Australia's publishing industry has some 1000 companies. Of these, approximately 150 are members of the Australian Publishers Association which represents the majority of the industry's total output. The majority of Australia's leading publishers are subsidiar-

---

[7] http://www.corporateinformation.com/data/ statusa/australiabooks.html

ies of UK and US firms. One of the larger Australian publishers is the Australian Government Publishing Service. State and Federal Governments produce 30% of the local book production. Of the 1800 booksellers in Australia, and some 2,500 general retail outlets also marketing books, 650 are members of the Australian Booksellers Association. In the late 1990s, export sales for books by Australian publishers represented 11.5% of total sales. Freight costs are a significant disadvantage for books produced on Australian paper for export. Government assistance to publishers is provided through various Departments and the Australia Council, and includes the reimbursement of authors and publishers for the loss of royalties due to the sale of books to libraries, and subsidies for the editing and production of literature.[8]

Growth in the printing and publishing industry's domestic and export markets is related to levels of education, income, gender and general economic outlook. That is to say that the tertiary educated have a greater propensity to read in comparison to people with only a primary education; those on higher incomes have greater disposable finances available for book purchases; and females tend to read more than males. Australia's involvement in the internationalisation of the market for education makes us only too aware of the expansion of tertiary education throughout many nations and the higher incomes these people have been able to attract. However, most are still educated in their vernacular.

'Corporate multiculturalism' (Mitchell, 1996) is now deeply implicated in attempts to produce meaningful, innovative developments in Australia's printing and publishing industry in ways that articulate its enmeshment and repositioning within the global multilingual knowledge economy. Moreover, corporate multiculturalism's focus on niche markets is part of continuing efforts to promote investment in the sustainability of the world's linguistic diversity. This is in a context where the political resistance to the project of globalising English is increasing the risks to business and shareholder investments. Being bound up with the global/local restructuring of Australia's printing and publishing, corporate multiculturalism is now being fashioned and harnessed to promote harmony around

---

[8] http://www.corporateinformation.com/data/statusa/australia books.html

linguistic diversity in the face of friction and animosity arising from the take-over of many languages by 'superpower' English.

Current debate within the industry is constructed as a struggle between the global open market and the protection of what is imagined to be a 'local' industry. It is as if somehow the local industry was in the past discrete from the global publishing and printing industry arising out of London or New York. This is not the case. The contemporary debate in the industry is between the legacies of past forms of globalisation, such as those associated with the Traditional Market Agreement, and the contemporary processes of globalisation that are affecting the restructuring and de-structuring of the Australian publishing and printing industry. Current investigations into the innovative potential of second-generation digital technology may be understood as an effort by the Australian printing and publishing industry, with the support of government and real world university research, to find ways to sustain its viability through re-positioning itself globally within the multilingual knowledge economy.

REFERENCES

Auh, T-S. (1999), Promoting multilingualism on the Internet: Korean Experience. (http://www.unesco.org/ webworldinfoethics2/eng/papers/paper_8.htm accessed March 5 2001).

Crystal, D. (2000), *Language Death*. Cambridge: Cambridge University Press.

Darwish, A. (2001), *The Translator's Guide* (Arabic). Patterson Lakes: Writescope.

Desktop Publishing, Microsoft Encarta. Copyright 1994, Microsoft Corporation, Funk & Wagnalls Corporation.

Dragona, A. and Handa, C. (2001), Xenes glosses: Literacy and cultural implications of the web for Greece, in G. Hawisher and C. Selfe (eds.) *Global Literacies and the World-wide Web*. London: Routledge.

Falk, R. (1999), *Predatory Globalization: A Critique*. Cambridge: Polity Press.

Griffin, M. (2001), Culture and imperialism: Amitav Ghosh says he feels in conflict when writing in English. *The Age (Extra)* Saturday, August 18, p. 5.

Kemp, S. (2001), Taking stock of your ethics. *The Age*, Saturday, August 4 (Business & Money), p. 5.

Kleper, Michael L. The Illustrated Handbook of Desktop Publishing and Typesetting, 2nd Edition. Windcrest Books, 1996.

Macedo, D. (2000), The colonialism of the English only movement, *Educational Researcher*, 29, 3, pp. 15-24.

Mitchell, K. (1996), In whose interest? Transnational capital and the production of multiculturalism in Canada, in R. Wilson and W. Dissanayake (eds.) *Global/Local: Cultural Production and the Transnational Imaginary*. Durham: Duke University Press.

Nettle, D. and Romaine, S. (2000), *Vanishing Voices: The Extinction of the World's Languages*. Oxford: Oxford University Press.

Ow, J. (2000), The revenge of the yellowfaced cyborg: The rape of digital geishas and the colonization of cyber-coolies in 3D realm's Shadow Warrior, in B. Kolko, L. Nakamura, and G. Rodman, (eds), *Race in Cyberspace*, New York: Routledge.

Reed-Scott, J. (1999), *Scholarship, Research Libraries, and Global Publishing*. New York: The Haworth Press.

Rose, J. (2001), *Ethical and Active Shareholding: An Australian Investor's Guide*. Elstern-wick: Wrightbooks.

Wallraff, B. (2000), What global language? Don't bet on the triumph of English. *The Atlantic Monthly*, 286, 5, pp. 52-66.

Warschauer, M. (2000), Language, identity and the Internet, in B. Kolko, L. Naka-mura, and G. Rodman, (eds), *Race in Cyberspace*, New York: Routledge.

Chapter 3

MULTILINGUAL BOOK PUBLISHING IN AUSTRALIA

Christopher Ziguras and Melissa Brown

This chapter provides an overview of the various ways in which Australian publishers have gone about producing books in languages other than English. While the rest of this book focuses on future developments, this chapter is deliberately backward looking, describing where we have come from so that we can make more sense of the new possibilities that information technologies may present. We make use of several brief case studies of local publishers that have been printing and publishing books in languages other than English in order to provide a useful account of the contemporary base on which innovative publishers can build to reach new audiences in new ways.

Throughout Australia's history, imported books have dominated the Australian market, and the majority of the larger publishers in Australia today are UK or US subsidiaries. During the 1990s, Australian publishers maintained a market share of more than 50 per cent (Ahern, 1997). In this context, Australian publishers have traditionally been focused on the local market, while multinationals with Australian subsidiaries have operated on a global scale. For this reason, until recently the Australian book publishing industry had been relatively import-oriented and inward looking. During the 1990s, many industries began to rethink Australia's relationship with other countries in the region, resulting in a widespread export push into Asian markets. As part of this process, Australia's publishing industry also explored ways of breaking into Asia. More recently, the need to develop an export focus has been highlighted by legislative changes allowing parallel importing that may make the industry vulnerable as long as it restricts its sights to the domestic market.

The Commonwealth government and industry organizations supported market research and export assistance schemes aimed at exploring and supporting market opportunities for Australian book publishers in Asia (Dowse, 1994; Dow, 1995). Australian publishers' business in Asia consists primarily of English language learning texts, children's books, academic titles, popular fiction and guide-books. Korea and other countries with large English-reading popula-tions such as Singapore, Malaysia and Hong Kong appear to be the most important markets for Australian publishers (Ross, 2001). The future of export markets for Australian books will be explored in greater depth in a subsequent report in this series that will focus on regional export opportunities for books.

For the purposes of this report, the most interesting aspect of the export focus over the last decade is the absence of significant discus-sions about multilingual book publishing, despite the fact that most of the books read in the region are not printed in the English lan-guage. Market research reports produced during the 1990s focussed on the export potential for English language books, but at the same time emphasised the talents of Australian translators and the coun-try's broad range of language skills (Dowse, 1994). By expanding the capacity to publish and print in languages other than English, Australia can expand its export markets while serving the needs of the world's readers who are increasingly being incorporated into a global knowledge society and need ready access to books published in their own languages. While the commercial dimensions of multi-lingual publishing will be considered in a future report, this report focuses on the technical capacity of the Australian book production industry to compete in other language book markets: issues of tech-nical development, cultural compliance, local business and legal practices and language expertise.

Printed books are now conceived, published and printed digi-tally, and increasingly the digital production of books takes elec-tronic files across national, linguistic and cultural boundaries. Tech-nological changes have made publishing books in languages other than English much simpler for Australian publishers and printers. Recent developments in desktop computer software and the devel-opment of a new, more inclusive, international character set (Uni-code) now allows different languages and scripts to be used inter-changeably on computers much more easily. In addition, the Inter-net allows publishers, editors, translators, proofreaders and printers in different parts of the world to collaborate more easily than ever before, so that translated books are able to be developed and distri-buted more rapidly and more efficiently. Alongside these develop-ments, the quality of translations produced by machine translation software, is improving rapidly. Currently, as previous chapters have described, these programs are adequate enough to provide readers with the 'gist' of a piece of text written in another language. As the quality of translation software improves, it is likely that translators will utilise machine translation to produce a rough version of translated text before final editing, rather than translating the text from scratch.

FROM MULTICULTURAL PUBLISHING TO MULTILINGUAL EXPORTING

In Australia there are millions of people who prefer to read in languages other than English. At the time of the 1996 census, about 2.5 million people (16 per cent of the population five years and over) spoke a language other than English at home and over 200 languages were spoken. The five leading languages other than English were Italian, Greek, Cantonese, Arabic/Lebanese and Vietnamese, each of which was spoken by more than 100,000 people. A further ten languages were spoken by more than 40,000 people. Greek, Italian and Arabic had the largest proportions of Australian-born speakers, partly reflecting a greater rate of maintenance of these languages among the second generation of these language groups, and earlier periods of migration (ABS, 2001). In the larger capital cities, local newspapers, magazines, radio stations and community television programs in many different community languages cater to speakers of other languages. A considerable amount of print material is produced in many different community languages in Australia, but little of this is in book form.

CASE STUDY: POLYPRINT

Established in 1960, Polyprint is a company that has built its reputation on its foreign language typesetting, design, printing and publishing capabilities. Predominantly a production house, Polyprint coordinates the translation of business documents, pamphlets, information materials, workplace documents, and training manuals into up to 120 languages. Polyprint's clients include government departments, tourism boards, universities and Australian offices of multinational organizations. Polyprint has a core team of ten full-time and four part-time staff, most of whom are bilingual or multilingual. Polyprint's own staff are able to deal with most small scale translation work into languages such as Japanese, Chinese, Greek and Arabic. In addition, Polyprint has a pool of up to 200 translators who are all accredited by the National Accreditation Authority for Translators and Interpreters. Currently, the languages most in demand for translation work include Chinese, Japanese, Arabic, Vietnamese, Korean, Thai, Croatian, Italian, Serbian, Bosnian, Spanish, Somali, Turkish and Macedonian.

The bulk of Polyprint's translation work is done off-premises by sending proof-read English language text as digital files to the translators. To ensure conformity in format, Polyprint and its translators use compatible operating systems. The Managing Director of Polyprint, Glenn Testro, believes that machine translation is not advanced enough to be useful to Polyprint, and is sceptical of machine translation technology being able to pick up the nuances and subtleties of language and ensuring cultural correctness in the foreseeable future. He concludes that:

> 'Even if machine translators get to a point where they are very accurate, we would only use them for the smallest and quickest of jobs, for example, the translation of business cards. The fact that a lot of the material that we translate is complex, technical and sometimes full of jargon, rules out the possibility of ever using machine translators.'

However, Polyprint is considering expanding its use of translation memory tools, which, through word recognition can cut down the translator's time by ensuring that updated sections of a large document are recognised and only the revised sections are translated. Polyprint has acquired some information about these tools and most of their translators currently use *Trados* software. Polyprint does not anticipate using translation memory software in-house, but if these tools prove successful and time-efficient, they would encourage their translators to purchase this or similar software.

In terms of its skill base, Polyprint could certainly be regarded as a multilingual and multi-skilled workplace. Most staff members are skilled in a number of areas, for example, writing and editing, small-scale translating, proofing, desktop publishing and designing and typesetting. Asked why Polyprint was not venturing more into multilingual publishing, Testro reiterated that Polyprint was first and foremost a production house, and that its primary focus was on desktop publishing, typesetting, and design. Despite Polyprint's involvement in multilingual publishing on behalf of other publishing companies, it has no plans to publish books in its own right.

Publishers and printers interviewed for the study were generally of the view that multilingual book publishing in Australia is not a viable commercial option as there is simply insufficient local demand for books in languages other than English. However, there has always been small-scale book publishing for local readers of other languages. Since the nineteenth century, migrant communities from non-English speaking countries have published newspapers, periodicals, and to a lesser extent, books, both locally and overseas to serve their own needs (Holmes, 2001). Most of the books published have been literary fiction for immigrant communities in Australia. For a closer look at multilingual publishing in Australia, see the *Bibliography of Australian Multicultural Writers* compiled by Sneja Gunew and others (1992).

CASE STUDY: ELLIKON

Ellikon Fine Printers, a mainstream printing company that has taken advantage of connections within an ethnic community to print, publish and export books in Greek, provides a useful illustration of diasporic printing and publishing possibilities. Managing Director of Ellikon, John Zapris, has drawn on his Greek cultural background and links to La Trobe University's Hellenic Research Centre to become a key Australian printer of Greek language books destined for Greek language readers in Australia, Greece and throughout the world. Longstanding connections within the Greek community have brought the company a considerable amount of Greek-language printing work, and Greek clients feel that a printing company with Greek lineage will take more care with Greek-language jobs than a printing company for whom the content was incomprehensible. Ellikon has also printed books in Japanese, Chinese and Italian, however its foreign language book printing constitutes less than two per cent of all production. There are no technical difficulties associated with such jobs as long as QuarkXpress files are delivered with fonts, and the growing use of the PDF file format has made foreign language quite straightforward.

As well as printing many Greek language books, in 2000 Ellikon published its first Greek language title, translated as *The History of Greeks in Australia*, by La Trobe University Professor Anastasios Tamis. Ellikon are now publishing an English translation of this book. The decision for this printing company to itself play the role of publisher in this case was not taken lightly, and Ellikon's managing director only went ahead with the project because of his intimate knowledge of the Greek-language book market in Australia and Greece. Despite not yet having been marketed or promoted heavily in Australia, the Greek language edition of *The History of Greeks in Australia* has already sold half of its original print-run of 2000, and will soon be launched in Greece.

In the future, electronic publishing and distribution of books is very likely to internationalise the book production industry, making it easier for publishers in Australia to publish books for readers of other languages both in Australia and overseas. Technological developments such as ebooks and printing on demand mean that readers anywhere in the world may be able to access large lists of titles, and many people around the world will be looking for titles in languages other than English. Because distribution mechanisms will change, it is important to look at what may become technically feasible in the future in order to assess how the international book market will develop. There is a tendency in the Australian industry to assume that the current production and consumption patterns will continue rather than imagining how they may develop and seeking to be at the forefront of those changes.

When thinking about publishing in languages other than English, a publisher of course first needs to think about where the readership for such books is. Just as English readers live all over the world in many different countries, readers of most other languages are also likely to be found spread around the world, both in that language group's 'home countries' and spread around the world in what are referred to as 'diasporic communities'. That is, the world's Vietnamese-speaking population is located primarily in Vietnam, but also in the United States, Australia, France and many other countries. At present it is very difficult to reach these audiences, because the market for books in Vietnam is small and the currency is weak, and because the Vietnamese reading community outside Vietnam is a very small and dispersed market segment. However, it is likely that 'diasporic public sphericules', as Stuart Cunningham (2001) has called them, will develop, linking Vietnamese communities in different countries through electronic media. This has happened through cable television stations that broadcast in Italian, Greek or Korean to different countries, and it may one day be possible for book publishing to operate through similar transnational public spaces that operate in languages other than English and link together people around the world who share a language.

CASE STUDY: COASIT

A good example of the interrelationship of diasporic communities and their book publishing needs is provided by CoAsIt's use of children's books. CoAsIt is a Melbourne-based community organization for Italians and Australians of Italian descent that provides educational programs to Italian language students from pre-school to adult learners. It generally uses imported Italian language texts but has had difficulty sourcing high quality children's books in Italian. Instead, the organization has translated many Australian and United States children's picture books into Italian. The translated text is laser printed and then pasted into the book over the English text. CoAsIt have written to the books' publishers, who have expressed no concerns about the process, as it results in sales for the publisher in any case. The lack of such texts in Italian raises the question of whether such titles would be commercially viable in Italy if published in Italian. An organization such as CoAsIt, whose staff have an extensive knowledge of the children's book publishing industry in Australia and Italy, is in a position to be able to translate such texts to suit the needs of Italian speaking children both in Italy and around the world.

Most of the books published in languages other than English in Australia were originally written in the language in which they were published rather than being translated from English, as was *The History of Greeks in Australia*. It has been rare for Australian publishers to translate books from English into other languages as the local market is too small to justify the cost of translation, and exporting has been difficult for these small publishers. However, with the digitisation of book production, publishing books in languages other than English in Australia, whether in the original language or translated from English, may become simpler and commercially viable. Currently, the primary means by which Australian books are translated is by a local publisher selling translation rights to an overseas publisher.

CASE STUDY: CIS

Heinemann One publisher that has some experience in exporting books in languages other than English is CIS Heinemann, the foreign language division of Reed Education Australia, a subsidiary of international publishers, Reed Elsevier. CIS Heinemann is Australia's largest publisher of texts for learning languages other than English, and produces books and associated media for primary, secondary and adult learners. Most of CIS Heinemann's books include English as well as the language being learned, and many of their books are exported without modification. Their Japanese secondary text series, *Kimono*, is the market leader in the United States. It is not economical for American publishers to develop their own texts for Japanese or Italian language learners so CIS Heinemann produces US editions of their books for a US publisher. These series, such as the popular *forza!*, are versioned into US English and adapted for an American audience by CIS Heinemann and all pre-print work is done in Australia before the books are printed in Singapore and shipped to the US. The European market has picked up the Italian course *Ci siamo*. To cater to European learners of Italian, CIS Heinemann have produced an edition of their *Ci siamo* series with only Italian text.

LICENSING TRANSLATION RIGHTS

At times, overseas publishers wishing to produce translated versions of their titles approach Australian publishers. Usually, the foreign publisher acquires the rights to publish a title in a particular language anywhere in the world. The Australian publisher usually sends an electronic copy of the text to the overseas publisher, who would then take full responsibility for translating, editing, publishing and distributing the title as their own. In return, the Australian publisher receives payment in the form of an advance, royalties and a split of income from subsidiary rights. Each country's rules and regulations only become a significant issue once an income is generated and the parties have to negotiate each country's tax regulations.

Because digital files can be sent around the world by email, there is now no technical impediment to such forms of translation, and parts of the book content that do not need to be translated, such as images and charts, can be kept in their original format in the translated electronic version of the book. For full colour books that are to be translated, the cost of translation is reduced if all text is printed in black, so that the same colour plates can be used for printing each language version.

GOVERNMENT ASSISTANCE

The Commonwealth government, through the Australia Council, provides grants to overseas book and magazine publishers to subsidise writers' and translators' fees and/or publication costs of overseas editions of the work of living Australian writers. The Australia Council awards up to $10,000 per title for works of fiction, literary non-fiction, poetry, drama, books for children and young people, anthologies of Australian creative writing and books of general cultural significance by, or about, living Australian writers (Australia Council, 2001). Australian publishers cannot apply for grants to assist with translation of Australian works into other languages but these subsidies do reduce the cost for overseas publishers considering translating some genres of Australian books.

During the 1990's, the Australia Council also ran a Translations Grants Program, which supported the translation of Australian books into languages other than English and of books in other languages into English. While designed to raise the profile of Australian translators, these grants also supported Australian authors in their attempts to break into foreign markets. For example, in 1996, well-known Melbourne-based Chinese poet, Dr Ouyang Yu received a grant to translate Christina Stead's novel *The Man Who Loved Children* into Mandarin Chinese and subsequently, the novel attracted attention amongst Chinese publishers. Although Australian publishers and literary agents were eligible for the grants, the majority of those who applied for the grants were translators. The program was discontinued in 1999.

OVERSEAS BRANCH OFFICES AND ALLIANCES

As described earlier, publishers in other countries who wish to publish translated versions of titles sometimes approach Australian publishers. Translation rights are usually negotiated and sold in this way on a case-by-case basis. If publishers believe there is a significant market for translated versions of their books in a particular market, they will sometimes form an alliance with an overseas publisher, who is able to translate, edit, publish, distribute and market their books. Instead of simply selling the translation rights, the two publishers work together to produce books that are branded by both publishers, and which are quality assured by both. In addition, such arrangements also involve reciprocal translations from the other language to English, and mutual distribution of each publisher's books in each country. If the Australian publisher wishes to retain full control of publishing overseas, they may establish a branch office in the target market, although this requires a more significant financial investment. While many Australian publishers operate as a branch of transnational publishers, it has been rare for Australian publishers to establish their own branch offices in other countries.

CASE STUDY: LONELY PLANET

Lonely Planet is of the few Australian publishers to have established overseas subsidiaries. In its 27–year history, Lonely Planet has grown from a modest enterprise into a global provider of up-to-date travel information. The Lonely Planet brand is recognised by independent travellers throughout the world, and appears on a huge range of travel guidebooks, travel literature, digital guides and electronic services. Lonely Planet has grown into a global enterprise with offices in Melbourne, Paris, London and Oakland employing nearly 550 people. Sales figures for Lonely Planet reflect its global profile – over 85 per cent of its sales are made outside Australia. In 2000–01, 47 per cent of sales were made in Britain and Europe, 33 per cent in the United States and Canada, 12 per cent in Australia and 8 per cent in Asia.

The French office, a wholly owned subsidiary of Lonely Planet Publications, was established in 1992 to launch Lonely Planet's first foreign-language publishing operation and handle sales and marketing in France. This office now publishes a list of more than 50 titles. French authors write a third of this list and other titles from Lonely Planet's English list are translated and customised for French-speaking communities across Europe, as well as in North America and Asia. All of these titles are produced in-house at the Paris office. Lonely Planet believes this local presence allows the company to connect with the local audience and to intuitively understand the local travel scene and its trends, as well as the information needs of French travellers. Lonely Planet also believed it would be much easier to develop a strong network of professional native French-speaking translators and authors who also share a passion and interest in travel based in France rather than in Australia.

Lonely Planet sold translation rights on an ad-hoc basis for many years. Their involvement consisted of negotiating an agreement, sending English-language digital files to the foreign publisher, and receiving royalties on the retail price. Lonely Planet guidebooks are currently produced under conventional rights arrangements in Bulgarian, Czech, German, Greek, Hebrew, Korean, Polish and Swedish. These books are published without Lonely Planet branding and with minimal editorial supervision or quality assurance by Lonely Planet. While Lonely Planet receives a reasonable income stream from licensing its content in this way, there are limited opportunities to creatively develop the market for travel content in other media, for example, digital forms, or to take full advantage of the brand recognition that has built up in many countries through its sales of English language editions.

For these reasons Lonely Planet has begun to enter into long-term partnerships with publishers with solid reputations and diverse business operations in the non-English language markets. Lonely Planet has recently formed new partnerships with Spanish and Italian publishers to make a range of Lonely Planet-branded products and services available in those markets. Although its partnership in Spain is relatively new, Lonely Planet's Spanish partner published ten Lonely Planet titles in the first six months of 2001. All production work for these co-branded books is done by the partner in the local market, with ongoing training, editorial and quality assurance input from Lonely Planet. While these books are currently printed separately in Spain and Italy and sold to the key Spanish and Italian markets worldwide, Lonely Planet is looking at ways to consolidate its European production, marketing and distribution aspects, such as printing its European-language titles through the one local European printer to achieve greater economies of scale. These new partnerships also incorporate physical and on-line distribution, digital publishing, content licensing and other business development functions. Lonely Planet's future strategy involves developing similar alliances in other key world languages such as Chinese and Japanese.

CONCLUSION

The scale of book publishing in languages other than English in Australia is quite small. Throughout Australia's history, books written in other languages have been printed locally for migrant communities in Australia but these books have generally not been exported to readers of these languages in other countries. Australian books written in English are usually published in other languages through the sale of translation rights to overseas publishers. Few Australian publishers have actively pursued exports to other language markets, although as the industry becomes more export oriented this may change. In order to export books in other languages, publishers must have access to high quality content that is in demand in that language and a network of commercial partners in these markets, as the Lonely Planet example illustrates.

In the past, the combination of national borders and language differences proved largely impenetrable for publishers. In the next chapter, we describe a number of technological developments that have made these barriers more permeable. Books, we argue, will cross both national and language barriers much more easily in the future, and possibilities will emerge for Australian publishers to become more engaged as providers of content in various forms to users regardless of location and in a variety of languages. As digitisation and the Internet change the ways in which books are written, produced, printed, distributed and read, Australian publishers will have to rethink the traditional means of translation and export described in this chapter in order to take advantage of new possibilities.

REFERENCES

Ahern, Annette (1997) Australia: Book Market Overview. Corporate Information.
Available at:
http://www.corporateinformation.com/data/statusa/australiabooks.html

Australia Council (2001) Literature Grants: Presentation and Promotion.
  http://www.ozco.gov.au/literature/present.html

Australian Bureau of Statistics (2001) *Year Book Australia 2001*. Canberra. Australian
  Bureau of Statistics. Also available at http://www.abs.gov.au/

Cunningham, Stuart (2001) Popular media as public 'sphericules' for diasporic communities, *International Journal of Cultural Studies*, 4(2) June 1, 2001 pp. 131-48.

Dow, Lesley (1995) *Australian Book Publishing Industry Market Research Report*. Canberra:
  Department of Communication and the Arts, Australian Book Publishers Association.

Dowse, Sara (1994) *Export Opportunities in Asia for Australian Publishing*. Canberra:
  Australian Government Publishing Service.

Gunew, S., Houbien, L., Karakotsas-Seda, A. and Mahyuddin, J. (1992) *A Bibliography
  of Australian Multicultural Writers*. Geelong: Deakin University Press.

Holmes, A. (2001) Case Study: Publishing in languages other than English. in M.
  Lyons and J. Arnold (eds) *History of the Book in Australia Volume Two: Towards A
  National Culture in a Colonised Market, 1890-1945*. St. Lucia: University of Queens-
  land Press.

Kirsop, W. (1998) Publishing in foreign languages in 19th century Australia: the
  French and German cases. Paper presented to the History of the Book in Australia (HOBA) Conference 1998, State Library of NSW.

Lawrence, Amanda (1998*) Japan-Australia Literature Report Part 2 Book Industry Research*.
  Parkville: Asialink.

Ross, John (2001) 'Doing Business in Asia', *Australian Bookseller and Publisher*. April
  pp.14-15.

ACKNOWLEDGMENTS

We would like to thank the following individuals and organizations for the assistance they provided in the research of this chapter: Megan Fraser, Lonely Planet, Liz Gasparini, CIS Heinemann, Michael Heyward, Text Media, Amanda Lawrence, Asialink, Glenn Testro, Polyprint and John Zapris, Ellikon Printing

Chapter 4

DIGITAL MULTILINGUAL BOOK PRODUCTION

Christopher Ziguras and Melissa Brown

The simple fact that books now exist as digital files before they are printed or read has profound implications which will be obvious to everyone in the printing and publishing industries. However, the linguistic dimensions of these technological developments receive little attention, especially in English-speaking countries. In this chapter we will consider the possibilities afforded by various technologies that allow for books to be more easily translated and printed in multiple language versions. We argue that new possibilities for multilingual publishing are created as a result of the various technologies that make books more mobile across geographic, national and language barriers. In particular, this chapter considers the use of electronic communications in manuscript preparation, the development of language-specific online bookstores, and the possibilities afforded by print-on-demand, dbooks and ebooks for Australian exporters of books in languages other than English. When pointing to new possibilities in this chapter, we are aware that many of the issues facing Australian exporters are commercial rather than technological, and these commercial aspects will be more fully developed in the next Module of the C-2-C project. In order to understand the new commercial environments in which publishing will operate on a global scale, it is important to understand the ways in which various technological developments will shape these emerging markets.

ELECTRONIC COMMUNICATIONS AND INTERNATIONAL PUBLISHING

Because digital files can be sent huge distances in seconds, there is greater scope for using a distributed network of people and services throughout the publishing and printing process. Publishers can more easily source particular services from lowest cost locations without this distribution of tasks slowing down the production process significantly. It is now common for 'Australian' books to be typeset in India and printed in China before being distributed in Australia. What are Australia's competitive advantages in such a global system? India has a large highly-trained workforce fluent in English in a country where wages are much lower than in other English-speaking countries. Southern China has low-cost, high-quality printing facilities that are thoroughly integrated into global publishing and distribution networks. It is up to each individual publisher to position themselves in a rapidly globalising environment, and given the diverse and specialised nature of the industry, many different approaches will be explored. In this book, we are emphasising the productive capacity of Australia's position as a predominantly English-speaking country with very well established trading, educational and cultural connections with the region and a large multicultural and multilingual workforce. With the growth of global English and the increasing integration of countries in our region into the global economy, there is an increasing demand for translation of books into English and from English into other languages. This is the case especially for technical, educational and scientific texts, which require highly educated writers, translators and editors. Electronic communications allow Australian publishers to draw on a highly educated multilingual workforce to provide translating and publishing services that are in high demand throughout the region.

CASE STUDY: BLACKWELL SCIENCE ASIA

Based in Melbourne, Blackwell Science Asia (BSA) is in some ways a typical transnational publisher, acting primarily as an importer and distributor of books published by the head office in Oxford, England, and publishing a small number of titles in Australia for the local market. As is the case in most multinationals, foreign language rights for BSA's titles are usually handled by the global head office. Despite this, BSA has developed a niche in translating and publishing medical texts and journals in China and Japan using teams of Australian and offshore staff connected by electronic communications.

In China and Japan, BSA has approached pharmaceutical companies to fund the translation and publishing of books and collections of journal articles previously published in English. BSA organises translation either by medical academics in China or a translation agency in Japan. Formatted Microsoft Word files are sent back and forward in English, Chinese and Japanese. The translated text is typeset and printed in Hong Kong and the finished books are delivered to the pharmaceutical companies in China and Japan who then distribute the books through their network of sales representatives. In each case, translation rights are acquired by BSA, who coordinates each stage of the process.

BSA also publishes several journals for Japanese and Chinese scholarly associations. Articles either in English, or in the local language and then translated in Australia, depending on the journal and the author. The journal's editorial committee in the home country makes all editorial decisions, while translating and copyediting is conducted in Australia.

These journals are published in English so that academic writers in China and Japan can participate in the development of their discipline, which operates on a global scale in English. The scholarly associations that publish the journals have sought an overseas publisher because BSA has access to a large pool of English-speaking science editors in Melbourne, and translation and editing services are cheaper in Australia than in the US or UK. In addition, the fact that Australia is in a similar time zone to these countries makes electronic communication more convenient. The journals are generally printed in China or Japan and distributed worldwide from there, and electronic versions of the journal are emailed to the United States to be put online by Blackwell's US subsidiary. BSA's relationships with scholarly associations and individual academics in China and Japan have allowed them to undertake other publishing ventures in these countries, building on an established track record and knowledge of the local academic community and academic publishing market. Initially, multilingual editors and project managers were employed by BSA, but as the relationship between the Australian office and the Chinese and Japanese partners became regularised, the ability to communicate in Chinese and Japanese became less important. BSA Australia's experience provides an indication of the type of mediating role that Australian publishers may be able to play, translating between English and national languages in the region, in order to allow local academic communities to participate more fully in the global development of their disciplines. None of this would be possible without email and affordable international telephony, but with the growth of these communications, such translations and international publishing ventures are increasingly necessary.

ONLINE BOOKSTORES

Online bookstores provide a convenient means for readers with Internet access to search for and purchase books that may be more difficult to find in a particular location. The large American online bookstores such as Amazon.com, Barnes and Noble and Borders Online are now being joined by online booksellers outside the United States that are able to offer lower delivery costs for nearby readers, more locally published titles and more books in languages other than English. Online booksellers make a wider range of books available to readers by centralising storage and inventory management, and reducing the overheads associated with physical bookstores. Online bookstores have the capacity to put all publishers on a more level playing field by reducing the costs of distribution for small publishers and allowing niche publishers and publications access to virtual shelf-space that would not be possible in a physical bookstore. Even self-publishing becomes possible through online selling. However, all publishers face the challenge of marketing books to prospective readers in ever-larger and more congested markets. The development of online bookstores in more and more languages allows for easy distribution of translated books to readers in different countries.

In its early stages, the Internet was dominated by English speakers, therefore the first online bookstores traded almost entirely in English language books. More recently, multilingual sites and online bookstores have developed. For example:

- **Amazon** now has several international sites that offer English language titles as well as books in Japanese, Spanish, German and French.[1]
- **Schoenhof's** offers the largest selection of foreign-language books in North America, including language-learning material for over 500 languages and dialects, as well as fiction, non-fiction, and children's books in 30 languages.[2]
- Greece's largest online bookstore, features English and Greek language texts.[3]
- **La Bancarella** based in Italy, sells books in Italian and a range of European languages.[4]
- **ChinaSite.com** lists a large number of Chinese language online bookstores based in Chinese speaking communities throughout the world.[5]

THE GLOBAL REACH OF DIGITAL BOOKS

The development of digital books has been plagued by imprecise and inconsistent terminology, which has caused much confusion. We use the term digital book to refer to an electronic file containing data that can be read as a book either on screen or after being printed. Most digital book formats can only be accessed, viewed and printed using particular types of software or equipment. For example, PDF files can only be created and viewed using Adobe software,

---

[1] www.amazon.com
[2] www.schoenhofs.com
[3] www.zevelekakis.gr
[4] www.bancarella.com
[5] www.chinasite.com/ECommerce/Bookstore.html

although the Adobe Acrobat Reader is freely available online and operates on all common computer operating systems.

Most digital book files are intended for printing, either by commercial printers or by individual readers. The capacity to store, send and print digital books easily has led to the print-on-demand phenomenon, in which some publications are printed in small print-runs as needed by the publisher, while others such as government reports are printed by individual readers on desktop printers.

Physical digital book reading appliances such as the Rocket eBook reader have appeared on the market recently.[6] These digital book-reading appliances are small screens that sometimes have a stylus interface and are able to hold several thousand pages of text. Digital books are loaded onto the reading device in a variety of ways. Some readers connect to a personal computer and require the user to download files from the Internet while there is a trend in the United States towards internal modems that allow the user to dial up and download a book file simply. While these digital book reading appliances have only sold in the tens of thousands of units worldwide, tens of millions of software digital book readers have been downloaded to personal computers. The most commonly used of these are Adobe Acrobat eBook Reader and Microsoft Reader. For the purpose of this report, we will follow Clifford Lynch in using the terms digital book and ebook interchangeably and distinguishing between 'digital book reading appliances' for specialized hardware devices and 'software book readers' for products that run on personal computers (Lynch, 2001).

Predictions of the size of the digital book and ebook market are hotly contested. Research firm IDC predicts that the U.S market for digital books, including digital downloads and print-on-demand books will grow from US$9 million in 2000 to US$414 million in 2004 (Enos, 2000). However the Ispos-NPD BookTrends survey found that 77 percent of online consumers were 'not very likely' or 'not at all likely' to buy an ebook (Nua Internet Surveys, 2001).

What is clear now is that, despite all the hype, apart from a few niche areas suited to ebooks, for example, textbooks, technical and reference books, most books will continue to be printed and read on paper regardless of how the electronic version of the book is pro-

---

[6] www.ebook-gemstar.com

duced and distributed (Tennant, 2000). Paper is still the preferred user interface.

While many of these developments are simply replicating the traditional book format in a digital environment, Clifford Lynch notes that the technology is allowing the evolution of new types of digital texts. In some ways, the use of the term 'digital book' disguises the fact that much of the digital text published electronically is not in book form. Reference texts such as dictionaries, encyclopaedias and manuals have become readily accepted in their digital form, precisely because they have always been searched and browsed as databases rather than being read in a linear fashion like a novel. These types of texts have been very successful in their digital form because they have changed to take advantage of the new media. The book is undergoing two different and distinct changes as it moves into the digital medium (Lynch, 2001).

In this book we have made a distinction between first generation and second generation software applications and operating systems in terms of their degree of internationalisation. Most first generation applications and operating systems were designed for particular environments and must be localised in order to work in different locations or with different languages. 'Localisation' involves adjustments to ensure that a particular local version will support the local language's character-set, rendering direction, hyphenation, number and time formatting, etc. Many of the new digital book production processes are part of a more globalised second generation, in that they have effectively eliminated the need for localisation by adopting global standards. For example, PDF files have all fonts included in the digital file itself so that a PDF file in any language can be read on any PDF reader on any computer irrespective of its language of origin.

Books can be distributed globally by traditional means, through having networks of publishers and distributors in different countries. However, increasingly, new technologies are allowing for new forms of distribution of books anywhere in the world in much smaller quantities. Below, we will discuss the development of dbooks, print-on-demand and ebooks, each of which makes books more readily available across borders to consumers. Before considering these new forms it is worth looking at the current state of national and global markets in books and various forms of digital content.

Publishing has always been constrained geographically by legal impediments to cross-border distribution. Historically British publishers had a monopoly on publishing in the colonies and later, the countries of the Commonwealth. The British cartel was supported by the 'British Publishers' Traditional Market Agreement' which divided the English speaking world in two, to be shared by British and American publishers. The British had Britain itself and the former British Empire while the Americans had the US and its dependencies. Publishers in one or the other had exclusive rights to their own territory (Richardson, 1996).

This has led to entrenched economic arrangements, which have arisen around these legal impediments and led many industry actors to be resistant to their removal. Recent industry opposition to the Australian government's deregulation to allow parallel importing of books from overseas even if a publisher has rights to the Australian market is a case in point (Fels, 2001; Nix, 2001).

As well as commercial reasons for restricting trade in books, many nations have selectively restricted the traffic in books (and in doing so have restricted the traffic in ideas) for political reasons. In recent years much controversy surrounded the realisation that copies of Hitler's *Mein Kampf* were being ordered through Web-based booksellers and then imported into Germany in contravention of that country's anti-Nazi statutes. As international book distribution increases, individual governments' capacity to regulate the reading habits of its population decreases. Of course customs controls can still intercept the movement of physical books across borders, and Web-based booksellers can be threatened with having access to their sites blocked in a particular country if they sell copies of books to buyers in countries where those books are illegal. When considering the global possibilities of digital books, it is important to bear in mind these entrenched commercial-legal barriers and the persistence of national censorship. On the part of network based information suppliers, there is now great interest in 'geo-location technologies' that allow online vendors to determine which nation users are based in. This allows them to tailor their services, and also to ensure that local laws are complied with (Lynch, 2001).

PRINT-ON-DEMAND, DBOOKS AND EBOOKS

While online booksellers usually distribute printed books, the Internet and the digitisation of book production also allows for digital files of books to be transported to and printed at the reader location. The most common form of this is online distribution of reports, books and various other documents as PDF files for readers to print on their own desktop or office printers. Many publications that were previously distributed without charge, such as government reports, are now distributed in this way from the organization's website. Research over the past few years has shown that readers will print most short articles and reports to read on paper, but prefer to have bound copies of books that will be read from cover to cover.

For books, print-on-demand has operated on a larger scale with the advent of digital book printing companies that print small quantities of books (1 to 1000) from PDF, QuarkXpress or Pagemaker files. One such company in Australia is dbooks, a subsidiary of one of Australia's largest printing and publishing companies, PMP Limited.[7] Dbooks allows publishers to digitally print small quantities as required or output the book as an online document or ebook file. Because PDF files are normally used, fonts appear the same way on every computer in all languages.

In the next few years, bookstores may have the capacity to print books for themselves. In the United States, a telecommunications company called Qwest ran advertisements in 1999–2000 suggesting that new fibre-optic networks would allow for 'every book, ever written, in any language' to be purchased in any book shop, and 'every movie ever made' would one day be available for pay-per-view at every motel (cited in Lynch, 2001). The technology for this process is still some time away, but is being trialed currently, and has the potential to radically change the way the industry operates (Epstein, 2001).

The development of printing on demand brings down the cost of printing small numbers of copies of books and reduces or eliminates the cost of transportation and storage. These are important to Australian publishers of books in languages other than English, as it makes publishing books of interest to geographically dispersed groups of readers viable. Such developments are likely to pro-

---

[7] www.dbooks.com.au

foundly change the economics of publishing books in languages other than English, especially educational and technical books for which there may be a large audience on a global scale but not significant enough demand in any one location to make translation, distribution and marketing cost-effective.

A survey of a number of ebook stores operating out of Australia and the US reveals that very few ebooks in languages other than English are currently available, though many of the large distributors are planning to stock foreign language ebooks in the future. The figures for the proportion of foreign language ebooks being published and sold are in dispute. One Australian-based ebook publisher, Zeus Publications, surveyed 30 ebook distributors and found that 3 per cent of ebooks available worldwide are in languages other than English. However, Kathryn Hardman from kdhbooks.com, a Melbourne-based ebook publisher, retailer and marketer estimates the proportion to be closer to ten per cent. Our survey of online ebook publishers and retailers revealed the following:

Company	URL	Languages available
Gemstar eBooks	www.ebookgemstar.com	English
Peanut Press	www.peanutpress.com	English
Barnes and Noble	www.ebooks.barnesandnoble.com	English
Electric ebook	www.electricebookpublishing.com	English/Spanish
ebooks.com	www.ebooks.com	English
Bookoo	www.bookoo.com.cn	Chinese
Libronauta	www.libronauta.com	Spanish
Ebooks Brasil	www.ebooksbrasil.com	Portuguese
Adlibris	www.adlibris.se	Swedish
Ebooks France	www.ebooksfrance.com	French
Cytales	www.cytale.com	French
Korea ebooks	www.koreaebook.co.kr	Korean
Zeus Publications Australia	www.zeuspublications.com	English, Russian, Greek, Croatian, Italian

Like print-on-demand, ebooks dramatically reduce the cost of distribution and the importance of location. As the technology develops, it will be taken up in other language markets as were the Internet and online bookstores, and in the near future it is likely that the majority of ebooks sold will be in languages other than English.

CASE STUDY: ZEUS PUBLICATIONS

Established by Bruce Rogers in 1999, Zeus Publications Australia (http://www.zeus-publications.com) was Australia's first ebook publisher and on line ebook store. Moreover, Zeus Publications is one of only a few electronic publishers with an English language website that actually lists and publishes books in languages other than English. Zeus has published ebooks in Russian, Greek, Croatian and Italian but foreign language books account for only around one per cent of sales. Rogers' main motivation for establishing Zeus Publications was to give the wider population the opportunity to read books from talented authors who may not under normal circumstances be able to have their book published, regardless of where they live and what their native language is. His overriding concern is providing quality fiction.

Zeus Publications does not actively market its foreign language titles; it simply lists them on its website as a service to the authors and to people who want to read foreign language books. The exception was Entelechia, a Greek language book that Rogers promoted in Greek and which is listed on Greek language ebook sites. Zeus' customers are predominantly from the United States, with Zeus starting to attract more interest from British and central European customers. Bruce Rogers believes it is only a matter of time and education before Australian audiences embrace ebooks. However, for the time being, the Australian market for ebooks will continue to be small as a result of our small population.

Zeus Publications has published five foreign language titles and expects to add additional foreign language titles to its website in the near future. There are currently no plans to translate any of the foreign language books into English. For the majority of people interested in downloading an ebook, there is no problem downloading a foreign language ebook. The book can be ordered in HTML, Word or PDF, and most computers support pan-European text, thus allowing users to upload the fonts required to read Russian, Greek and Croatian text. Zeus Publications' role as go-between for non-English speaking authors and non-English speaking book-lovers is interesting Although current interest in foreign language titles is coming predominantly from the United States, there is no reason why Zeus Publications will not eventually cater to the needs of multilingual book lovers in Australia.

CONCLUSION

With these emerging book forms, access to global networks of information and cultural exchange becomes more important than geographical location. And while national boundaries becomes less important, being able to negotiate and move between different language markets (rather than national markets) will become more important in disseminating intellectual and cultural production around the world.

The benefits of international digital book publishing lie in the local publisher retaining full control of the production process without having to involve any intermediaries. An English language text can be emailed to translators, proofreaders and typesetters located either in Australia or overseas, and can then be published electronically from Australia for global access or sent to a printer anywhere in the world. At each stage, the publisher is able to retain control of the process, as long as they have a trusted network of bilingual staff. The disadvantages are that without a commercial presence in the target market, it is more difficult to market and distribute one's books to a distant readership. As the technological possibilities for book publishing and distribution in such a global and multilingual knowledge society develop, the question of how to market books in such an environment becomes a key issue. The commercial implications of these technological developments remain to be seen, but we are confident that Australian publishers can play an important role in the future of multilingual publishing.

REFERENCES

Enos, L (2000) Report: E-book Industry Set to Explode. *Ecommerce Times.* 20 December. Available at www.EcommerceTimes.com

Epstein, Jason (2001) Reading the digital future. *The Australian.* 4 July. pp.34-36.

Fels, Allan (2001) Publishers be damned, readers deserve a fair go. *The Age,* Melbourne. 14 July, p.7.

Lynch, Clifford (2001) The battle to define the future of the book in the digital world. *First Monday.* 6(6) Available at:http://firstmonday.org/issues/issue6_6/lynch/index.html

Nix, G. (2001) I could write a book about Australian authors, publishers and deaf ears. *Sydney Morning Herald.* 4 June.

Nua Internet Surveys (2001). Bright future of ebooks seems far away. Nua Internet Surveys. Available at http://www.nua.ie/surveys/index.cgi?f=VS&art_id=905356831&rel=true

Richardson, D. (1996) Copyright and Monopoly Profits: Books, Records, and Software. *Current Issues Brief 15, 1996-97.* Department of the Parliamentary Library. Available at http://www.aph.gov.au

Schofield, J. (2000) Net growth faces language barrier. *The Age.* 4 July.

Takuzo, T. (2001) Transmart, *Nikkei Net Business.* No. 70, pp.94-95.

Tennant, Roy (2000) The emerging role of e-books. *Australian Bookseller and Publisher.* November 2000

ACKNOWLEDGMENTS

We would like to thank the following individuals and organizations for the assistance they provided in the research of this chapter: Liz Gasparini, CIS Heinemann; Kathryn Hardman, kdhBooks.com; Mark Robertson, Blackwell Science Asia and Bruce Rogers, Zeus Publications.

Chapter 5

TYPESETTING MANUSCRIPTS

Gus Gollings

This chapter, and the following chapter, examine typsetting and computerised character encoding as a context for discussing issues that arise for printing and publishing in languages other than English and scripts other than roman.

Typesetting today is evolving into an automated process that follows general rules. This is an environment in which written communications – printed books and ebooks for instance – are more easily produced in a variety of languages (even to the extent of mixing languages on the same page), than ever before. The make-up and history of today's publishing environment is shaped by ancient human language practice on the one hand, and on the other, the technological fronts of standardised text encoding and mark-up. The use of a fixed, paginated or binary textual mark-up is often unnecessarily inflexible in a networked environment wishing to offer support to a variety of platforms and devices that operate across a variety of languages. This chapter traces the habits of typesetting over a millennium to better understand the technical drivers behind shifting trade practice as we witness the maturation of the file formats and encoding schemes that make multilingual text production more accessible.

IN THE BEGINNING

The earliest dated example of typesetting is from a hand printed scroll of the Buddhist text, titled in English, *the Diamond Sutra,* from AD 686 (or 1315 years before this publication).[1]

The Diamond Sutra was made through a process of first drawing on paper the design of each section of ideographs, or imagery, in liquid ink. Then, the paper would be laid, ink-side down, on the levelled face of a wooden block, so as to leave a reversed impression

---

[1] http://www.thebritishmuseum.ac.uk

of the hand drawn design. The wooden block would then be carved in relief, leaving characters raised from the block face. Making a print was a simple matter of painting ink onto the carved block-face, laying parchment over the block and rubbing the parchment to make an impression.[2] This process is known as xylography. It is a precursor to the idea of movable type, where the unit of type is broken down from an entire 'block' of text, to the individual characters of the text. The *work* of typesetting found in this example (the drawing and carving of Han ideographs into blocks) concerns itself with the same fundamental constraints that typesetters face today. From *Figure 1* below, it can be seen that over a millennium ago, the vulgar elements of typesetting: text layout; font size; font selection and the placement of imagery, had all been methodically negotiated and solved, to a greater or lesser degree, and can be seen as legitimate components of the modern typesetting form.

Figure 1: A detail from the Diamond Sutra scroll

Suffice to say, at least thirteen hundred years ago, the basic procedures and concerns of typesetting were well established. Aesthetics aside, it is the productivity of various typesetting techniques that determines their selection and adoption. Since the printing of *The Diamond Sutra* there have only been three great leaps in the productivity of typesetting technology: Gutenberg (movable type), photo-

---

[2] Ibid.

composition (scanners and imagesetters) and the third is upon us presently and lies in the realm of programmable, abstracted and rule-based automation of character composed texts.

The following section discusses variations in the practice of setting type between 700AD and 1970, including movable type, its mechanical automation and the introduction of the keyboard. This leads into a more detailed outline of the period after 1970, in which mainframe computers and imagesetters are found. It is this period beyond 1970 that is taken as the beginning of the digital revolution proper. The impact of the digital revolution on typesetting practices and methodologies is not a straightforward innovation or 'natural' progression from the mechanical age to the digital age, as is the case in the late medieval transition from manual to mechanical typesetting and printing processes. It is found that the digital revolution has imposed a new paradigm on the preparation of arranged text that alienates it from the physical printing processes. Accommodating multiple delivery channels from the self-same source file through metadata-imposed transformation layers and style sheets is the new game of typesetting.

SMALL IMPROVEMENTS TO THE TYPESETTING PROCESS

The xylographic process described above takes the wooden block (containing many characters) as its smallest movable element, rendering it impossible to edit the character positions, spacing and size once the carving is complete. The transition to the next phase of typesetting logically entails splitting the wooden block into its component characters. This allows for rapid assembly of printing plates and division of labour along lines of character creation (drawing and carving of individual characters) and character composition (arranging characters as a text). Here, the early separation of font creation (calligraphy) and typography (the art of typesetting) is apparent, and the smallest unit of the composed text-to-be-printed was a single character or letter.

Movable type was used in China in the 11th Century but at that time production-oriented typesetting was not realised in an efficient form. The logic of moveable type posed a difficult challenge to Chinese ideographic typography because of the many tens of thousands of characters needed. The use of type, for the while, remained a laborious process. However, when moveable type was adapted for the European Gutenberg printing system efficiency and quality coalesced, providing a solid framework for the expansion of publishing networks. Alphabetised languages do not require the extended labour of character based language mapping, and therefore are more efficiently encoded.

Mechanised type-stacking was popularised in 1822, when William Church patented the first typesetting machine, complete with the first keyboard for the selection of type. There is a remarkable similarity between the concepts used in Church's typesetting machine and the concepts used in the early computer file-systems of the late 1960s and early 1970s. These computer file-systems were, of course, the beginnings of POSIX, or today's Unix operating systems. The similarity with the Church typesetter is seen in a) the use of a keyboard to input (select) a sequence of characters, and b) the limited capacity that handled only one 'line' at a time. This similarity goes some of way towards explaining the common ancestry that the printing industry shares with the computer world. The same division of powers was operating in the one, as in the other – input, storage, transport of data and, eventually, execution.

Each line of text produced by the early-nineteenth century typesetting machines was constituted by a 'stack' of lead type, one character sitting after the next, held together by ligatures. The typeset 'stack' would be chopped-up into lines of equal size and arranged into justified columns by hand. The obvious extension was to enable the mechanical setters to drop type on multiple lines of a page in the one session. Approaching the twentieth century, a multi-line editing feature was added, as well as the ability to key the line of type in the natural sequence, as opposed to backwards (as individual characters originally had to be set in reverse, 'wrong-reading', order so the print they made would be 'right-reading.' Interestingly, in practice the automatic line-setting of type on multi-line machines was no faster than doing it by hand one line at a time, for each typed line would have to be manually justified all the same. The problem of automatically justifying hadn't been solved adequately and columns of text could only come off the type selection machines left aligned or 'ragged-right'. An interesting aside to this early justification problem, is that the exact same problem is faced in the presentation of HTML pages containing justified text today.

In America, during 1880, Ottmar Mergenthaler conceived of *Linotype*, the famous typecasting compositor, bringing significant speed increases to the work of typesetting. A doubling in the speed of the *Linotype* systems was a seen in Tolbert Lanston's system of *Monotype*, developed just five years later in 1885. More importantly, the introduction of *Monotype* saw the breakdown of the typesetting machine into two separate components of keyboard and castor (a machine that arranged lead type into words, lines and columns), complete with a protocol (recorded on perforated tape) that allowed the keyboard to store its input (the sequence of keystrokes) for processing at a later stage by the castor. And the other way around, the castor could be kept at full speed by feeding it the completed 'jobs' (lengths of coded perforations in tape) from several keyboards operating in series.

The streamlining of the *Monotype* systems by 1930 saw productivity levels double yet again, reaching an impressive 20,000 characters per hour from a single castor. Further improvements were realised with the teletype typesetting systems that provided more automation than that of the basic Monotype machine. With teletype machines, the project of automatic justification was again attempted, and this time solved to some degree of satisfaction by engineering a contraption to read the perforated tape strips and make judgements about where to break the text up and where to insert spaces so as to justify the text in columns. Parallel developments in the setting of type in scripts other than Roman also benefited from teletype justification, as electrical technology is always language-independent, in that what it provides is a manipulation of symbols which can, but not necessarily, be designated as part of written language.

Here we come to a juncture in the history of typesetting, where several core attributes of the computer are harnessed to manage the composition of text (or, less popularly, several core attributes of the typesetting machines were borrowed to make a computer):

- *Algorithm based analysis* of character sequences leading to the most obvious developments of spell-checking, thesauri, etc. and also more profound parsing routines.
- *Metadata handling* of logical rules applied to, and saved with, sequences of characters that castor-type devices could detect as instructions to perform functions other than just the placement of type end-on-end.
- *'Saving' and recomposition* of entire texts and components of text that have previously been input.

- *Modem-transfer* of the sequenced and 'marked-up' text code over telegraphic wires (in a similar way to the transmission of Morse-code) between two computers.

THE RISE AND RISE OF DESKTOP PUBLISHING

The origins of typesetting are to be found in the art of hand-crafted book making that produced the fabulously decorated religious texts from the Byzantine Empire, the Middle Ages and the Renaissance. A series of combining forces centring on the need for efficiency in the typographic processes replaced the laborious art of the scribe with the technical skill of the typesetter. With the typesetter arose a new art, not entirely lost in patient subservience to machines. This art involved the design of cast metal fonts – Aldus, Baskerville, Bodoni, Goudy to name a few – and the beautiful set out books which exemplified the designer's art. With computerisation, the *art* of typesetting now is in thinking creatively about new information horizons, whilst building on the solid foundation of historical tradition.

As we have seen, technological developments in the typesetting trade during the millennia before 1970 are characterised by their linear development pattern, each moment doing its best to perfect the last, with flashes of new invention. From the 1970s, computer technology and its effect on typesetting practices became more and more dramatic. Typographic practice eventually broke down into three main groups: WYSIWYG (What You See Is What You Get) desktop publishers, career professionals and computer scientists. These groupings are not customary, and require further explanation:

- WYSIWYG desktop publishers are the vast majority of people using computers. They use Microsoft *Office* products and various downloaded shareware under Microsoft *Windows*;
- Career professionals are the people who *used* to be typesetters proper, or typographers. They are commonly known as graphic designers in Australia, and while they do use desktop publishing tools, they tend to operate on a different level to the WYSIWYG user;
- Computer scientists, which is an umbrella terms for hackers, 'sys admins,' computer scientists, engineers and others who survive on the food of open source computing and Unix environments.

WYSIWYG users	Professionals	Computer scientists
Microsoft *Office*	*Freehand*	TEX
Shareware editors	*Photoshop*	L^ATEX
	Fontographer	MetaFont
	QuarkXPress	Postscript
	FrameMaker	DocBook
	PageMaker	XML
	Illustrator	*et al*

Table 1: The tools of the three broad groups of modern typesetting

The three groups shown in Table 1 hold the reigns of typesetting today. The social context of technological change, the 'mainstream revolution' here, has meant that those with the least skill in dealing with the vicissitudes of typography (the left hand side column) produce the most typeset material and the professional users of the of the mainstream page setting, design and layout programs (the middle column) are perennially frustrated by the 'niceties' of their desktop environments which ensure they don't have access to the heart of the process. The remaining, right hand side, column is where the state-of-the-art is defined and it is where the art is perfected before its dissemination through mainstream (WYSIWYG) software. Trade and professional typographers in the past were those people designing and inventing new ways of producing printed material. In our present situation, the brains have been separated from the arms, as it were. Imagine for a moment if Gutenberg was able to arrange words on a page, but was unable to affect the process by which they were laid there.

There is an even greater gap between *what you get* and *what you mean* in the word processing of languages that do not use an alphabetised script. Character input of such ideographic script is a problem that is not entirely solved, although there are several general approaches. The most common is to use a romanisation system, like Pin-Yin for Chinese, so that one can type the *sound* of the ideographic character phonetically in Latin script, and have the computer select, on your behalf, an appropriate ideographic character (or sub-set of characters for further user-selection). The frustration of this process is knowing that there is a character in the character set that you want to input into your document, but having no way of 'keying' that character.

The next important stage of the digital revolution is to reconfigure the alphabet soup it has created in the streams of users outlined above. Humpty-Dumpty needs to be put back together again, and in such a way as to create a level playing field for creators, managers and users of typeset material. This solution must successfully negotiate issues of the open-source/proprietary dichotomy, global standards (in language and operability) and provide a coherent set of tools that share a common logic and operative measure which guarantee both the quality of trade practice established in scientific laboratories and equal access to those tools by all potential creators of text.

TYPEWRITER TO DESKTOP COMPUTER

Early typewriters began to appear in the 1880s, using the familiar QWERTY keyboard layout developed to eradicate jamming caused by the collision of the impression-making hammers. The idea of the QWERTY keyboard was to separate the most common letter pairs onto either side of the keyboard, in order to minimise the frequency of jamming hammers. Unfortunately, the QWERTY system is not the most ergonomic layout and has remained with us as a small proof that standards are more important to users than the 'best' solution to a problem. Once a technology has been introduced, learnt and accepted by a community it is hard retract it.[3]

In 1933, IBM bought Electromatic Typewriters Inc., of Rochester, New York. This acquisition not only allowed IBM to enter the typewriter business, it also provided the company with additional technical expertise that could be applied to its tabulator printers. IBM inventors foresaw this crossover application of typewriter technology as early as 1928.[4]

For the next fifty years IBM continued to develop electric typewriters, including in 1946, an electric Chinese ideographic character typewriter with 5,400 ideographic characters available via a cylinder. IBM began to take the first steps toward computing technologies in 1944, building what were essentially large calculators for accounting purposes. As these large calculators continued to develop into the 1950s they began to be marketed to the scientific community who, through government and institutional support were more likely to be able to afford the product and enable its continued development. The electric typewriter, on the other hand, remained the staple of

---

[3] A contemporary example of this feature of human interaction with technology is seen in the slow adoption of new internet technologies beyond HTML 4.
[4] www.ibm.com.

the workplace. By 1960, IBM was marketing specifically along two lines – office products (including smaller versions of their most complex computer systems scaled down into accounting applications, and electronic typewriters) and mainframe solutions for government army and large business.

Changeable fonts and inbuilt erasing strokes gave users of the IBM 'Selectric' Typewriters their first taste of typesetting customisation. The horizon rapidly changed coming into the 1980s with the US$7,895 'Displaywriter', marketed as a low-cost, user-friendly typesetting tool, having the added benefit of storage onto disk and spell-checking. Already this system was just a stone's throw away from today's personal computer (PC), having a display, keyboard, logic unit, 224K bytes of memory (in the high-end configuration), printer and a diskette read/writer. The 'Displaywriter' was superseded in July 1981 by the System/23 'Datamaster', a combination of both word processing and data manipulation – sold by IBM as an 'information processing' unit. Two users could operate it at once, provided, as the old joke goes, that they didn't use it at the same time.

In 1981, the office-oriented product line offered by IBM was re-launched on a new, singular platform. That platform was the first PC, complete with DOS version 1, word processing software, accounting software and a dot-matrix printer.

a   *b*   *c*   *d*

Figure 2: from left – a) 'Selectric' Typewriter, b) Displaywriter,
b) System/23 Datamaster, c) The first Personal Computer from IBM, 1981.

These developments in typewriters and computers directly fed into typesetting technology, the first part of the printing process to involve computer mediation. Hot metal typesetting systems such as Lynotype and Monotype, in their final and most sophisticated form, were based on a mechanical computational system – text entered onto a QWERTY keyboard punched holes in tape, which was then fed into a machine which calculated line justification and output hot metal type, line by line. Computerisation came in the late 1970s, replacing three key aspects of this technology. First, text was entered through an elementary word processing system, with the operator using control codes to mark-up point size, paragraph breaks and the like, whilst the computer calculated spacing for justi-fication and determined suitable line breaks. Second, data storage was electronic, saved to eight and a half inch discs. Third, images where 'photoset' from disc. In the first generation of these techolo-gies, a flash of light was projected from the inside of a spinning drum wrapped in a negative character set, onto a roll of bromide paper producing continuous galley. Optical devices determined the font size according to the mark-up instructions. In a second genera-tion of these technologies, the light source for character-by-character projection was a cathode ray tube, a television monitor if you like, projecting one character at a time onto bromide paper, massively reduced (and thus quality enhanced) by the optical projection sys-tem. These hybrid digital-analogue technologies, despite all their limitations, paved the way for fully digitised page creation and ren-dering, capable of dealing with texts and images together, and much greater efficiency and design flexibility.

POSTSCRIPT AND THE DESKTOP COMPUTER

In 1984, PostScript was released. PostScript is a programming lan-guage optimised for printing graphics and text (whether to paper, film, or computer screen is immaterial). In the vernacular of the day it was a 'page description language.' It first appeared as a commercial product in the Apple LaserWriter in 1985, licensed from Adobe (an offshoot from Xerox) where its strength as a device independent language (so that it could be sent to and processed by a variety de-vices) was quickly acknowledged and an industry of developers and users started to form around this interoperability.

Due to its complexity, PostScript required a powerful (by the standards of the day) processing system to run on. During the first years of its existence, the PostScript printers themselves had more processing power than the computers interfacing with them. The cost of the processing power was offset by the usefulness and pro-ductivity gains offered:

- PostScript is device independent. On a laser printer, 300 dpi output is achieved, while the same file gives you 2400 dpi out-put on an image-setter.

- To computer users, PostScript meant their software would print to any printer that supported the language. Users could choose the output device best suited to their purposes.
- Manufacturers could buy a license for the PostScript interpreter and use it to build an output device that would, by way of the growing popularity of the standard, be a relevant commercial pursuit.
- The specifications (syntax) of PostScript are freely available, and programmers are encouraged, if not required by the weight of demand for compliance, to build software that supports it.

POSTSCRIPT'S WIDESPREAD ADOPTION

PostScript was a high-risk venture for Adobe, taking 20 person-years to engineer (actual development duration of version one was a little over a year). Adobe received support for PostScript from Apple Computer, Inc. Steven Jobs, then Chief Executive Officer of Apple Computer, oversaw a US$2.5 million investment in Adobe's technology and commissioned a PostScript controller for the Apple LaserWriter printer. The PostScript controller for the LaserWriter printer allowed it to produce image-setter quality text and graphics.

At the same time, Aldus software was a start-up company and one of their first applications, PageMaker, was to establish a golden triangle with Apple's hardware and printer and Adobe's PostScript. Desktop publishing was born and within a year the combination of the LaserWriter, PostScript and PageMaker became a landmark. A major graphic arts supplier, Linotype, recognised the value of Post-Script and developed an imagesetter with its own PostScript imaging engine. Other manufacturers soon created business models to compete with Linotype, and PostScript, being at the centre of this new market opportunity, quickly became the *lingua franca* of the prepress industry.

Into the twenty-first century, prepress bureaus continue to make a living from the same core PostScript technology. The Portable Document Format (PDF) leverages from the same principles and practice distilled in PostScript and marks the only significant development to PostScript in over a decade and a half.

THE SLOWER AND SMARTER TYPESETTING REVOLUTION

In the world of scientists, mathematicians, and others who need to produce high-quality, aesthetically pleasing text, especially where technical content is included, there is no other option but to use the most advanced typesetting system known as TeX (including its myriad derivative tools and macros) initially created by one of the greatest computer programmers to have lived, Donald E. Knuth, and subsequently extended by TeX users from all over the world (http://tug.org).

The TeX software has now reached a level of maturity that few pieces of software have ever been able to achieve. Many people (designers and scientists alike) have celebrated Knuth's monumental coupling of programming and typography. Popularised in 1986, improvements to the systems have followed developments in digital printing technology and reflect corrections submitted over the years by thousands of volunteers.

The important lesson to take from the TeX typesetting system is framed by the mantra of modern text manipulation 'separate presentation from content'. This simple phrase wields a powerful concept that has now become the backbone of metadata and textual mark-up philosophies alike. Instead of telling the word processing application to use a particular font at a particular point size when the user is creating the text, the user should leave such 'presentation' information out of the text creation process, and focus instead on the words on the page. The words on the page need to be structured. However, it is important to make a distinction between structure and presentation. By structure we mean 'tagging' headings, paragraphs, block quotes, bulleted lists, index terms, acronyms, etc. so they can be identified as distinct *parts* of the one coherent flow of writing. By presentation we refer to the stylistic attributes that are given to the 'tagged' parts of the text. The magic of separating presentation and content is that one can rely on the *content* document as a definitive master source of the text which can be transformed by a presentation layer into almost any other rendition of the master text. A rendition of the master text could take the form of a PDF file, or an HTML website, or a Microsoft Word document, or even an audio file.

Content marked up in such a way as to exemplify the concept bequeathed to us by Donald Knuth, *separating content from presenta-*

tion, is more open and therefore easier to archive and leverage from than content contained in any other format. In comparison to using Microsoft Word to input your text, where you are locked into Microsoft's proprietary binary format, marked-up content can be seen as device independent and format independent. This principle is slowly filtering its way into mainstream computer use.

ADVANCED SOLUTIONS LEAD TO NEW PATHWAYS

Any recognition of the similarities between the very first and the very latest typesetting regimes, should not disguise or confuse the foundational shift in the constituent modality of typesetting we are witnessing today as a result of the digitisation of text. That is, recent changes in typesetting, brought on by the stellar growth of net-worked book production environments, are redefining the scope, effect, and positioning of typesetting within the book production chain. *Scope*, because advances in the area of text 'granularisation' and 'internationalisation' are changing the way books are composed, decomposed and recomposed by interactive and automated systems. *Effect*, because the proliferation of 'flexible' text delivery devices (from wireless web-browsers to speech synthesisers) has made the digital representational-integrity of the typeset page uncertain. *Positioning*, because typesetting is becoming more an issue of templates, rather than specific text-treatments; typesetting no longer needs to be fixed to individual books (rather books are fixed to it by way of style-sheets); and by extension, one no longer has typeset books, one has typeset templates which are temporally detached from the book production process and the books themselves, and as such typeset-ting is no longer subject to the customary production deadlines.

The separation of presentation and content into two separate entities (that nonetheless interoperate) has become one of the leit-motifs of text processing in the twenty-first century. We thought the digital era of typesetting was called 'prepress' just a few years ago, now the task of typesetting in the second phase of the digital revolu-tion is more is challenged by shifting relationships betweens texts, renditions of those texts and output mediums that at the best of times are *multiple* (electronic book and printed book) and at the worst of times *indeterminate* (consumer defined and open ended). The continuing evolution of typesetting standards, protocols and labour divisions are elusive as we move into publishing environments that demand accommodation for not only all language scripts (which we discuss in the following chapter) but also all output possibilities.

REFERENCES

Desktop Publishing, Microsoft Encarta. Copyright 1994, Microsoft Corporation, Funk & Wagnalls Corporation.

Kleper, Michael L. *The Illustrated Handbook of Desktop Publishing and Typesetting,* 2nd Edition. Windcrest Books, 1996.

Chapter 6

MULTILINGUAL SCRIPT ENCODING

Gus Gollings

If computers across the world are networked, then there is a need for a common computing platform (regardless of Windows/Macintosh/Unix differentiation). By 'common platform' it is meant that stored files from one computer must be able to be read and filtered through identical semantic processes on any other computer. There are many different computers all around the world today, using many different semantic processes. Because of this, it is easier to suggest a unified semantic system to which all computers can conform and thus interoperate, than it is to provide mechanisms through which one system can be converted to another, and so on. Work on the unified semantic process suggested above is already underway and has several incarnations, the latest being the much-debated character encoding system, Unicode.

Computers are fairly simple on one hand, and fairly complex on the other. They are simple because they can only do two things that any human can do routinely, that is calculate binary arithmetic operations and perform Boolean (true/false) operations. The complexity of the computer comes from the systems that are built around the two simple operations that a computer can perform. These systems are myriad and built by increments. Therefore, the complexity of a computer is seen to become greater over time, when in reality computers are still as simple as they always were: adding, subtracting and testing for truth. The use of computers in different parts of the world by different vendors and communities has led to the development of a diverse mixture of symbologies and systems to tackle the various problems that local scripts, sorting systems and typographical conventions entail.

As was detailed in the previous chapter, the early 1980s saw the calculator and the typewriter combine to form the personal computer (PC). The PC can do two things, calculate and word process. In reality, the calculator grew to encompass not only the spectrum of numbers, but also the spectrum of characters belonging to the particular language script that it was configured with. It did this by assigning numerical values to each of the characters (A, B, C, and so on, for English) that it wanted to represent and store. This assignment of numbers to represent characters was originally called a character code, and was developed to simplify the telegraphic transfer of messages over the 'wire.'

FOUNDATION OF TEXT PROCESSING

We can point at three developments that each offered notional solutions to a whole range of issues that are now addressed by the modern PC. The developments are:

- nineteenth century printing technology;
- weaving technology; and
- telegraphic communications technology.

Printing technology gave rise to keyboards and the decomposition of books into individual characters (a unit of type). Weaving technology supplied the idea of storing data on a punch card (the earliest form of memory in the sense of computer memory) and the idea that a pattern weaved in fabric is analogous to a pattern weaved in words and numbers. The telegraphic communications technology gave a sense of the need for efficiency of text processing in a commercial environment (time is money) and it also helped to structure the systems by which text would be encoded as dots and dashes (or zeros and ones, binary numbers).

Binary numbers are the same as our normal numbers, except they are expressed in base 2, as opposed to how they are normally expressed in commerce and day-to-day life in base 10. The following table will help to demonstrate the equivalency:

Binary numbers are the only numbers that can be stored in a computer's memory. Memory is literally a sequence of ons and offs (zeros/ones, charged cells, not-charged cells). Memory can

Figure 1: Decimal and Binary equivalency.

Base 10	Base 2
0	0
1	1
2	10
3	11
4	100
5	101
6	110
7	111
8	1000
9	1001
10	1010

be stored on hard disk or in Random Access Memory (RAM). Therefore, a computer only knows how to handle text by first converting it to a binary number that it can store and manipulate. A computer even draws typed numbers to the screen by first assigning them to a binary number, just the same as text.

From the table (*Figure 2*, right), a computer working with the 7-bit ASCII character set (explained below) would store the letter 'A' by writing to memory the binary number '1000001'. Conversely, when you type the letter 'A' on your keyboard the computer receives the binary sequence '1000001' and it then goes off the character assignment table and matches that binary sequence to the letter 'A' which is in turn mapped to a graphi-

Figure 2: An example of binary number assignments to characters on the screen in 7-bit ASCII code.

Character on the screen	Binary value used to process it
1	0110001
2	0110010
3	0110011
4	0110100
5	0110101
A	1000001
B	1000010
C	1000011
D	1000100
E	1000101

cal representation of the letter 'A' in a font file which then gets drawn to the screen.

Text manipulation on a computer offers a great deal of flexibility over the conventional methods of handwriting or using a typewriter. Digitised text can be stored in a file and transmitted to other computers for further processing, archiving, sorting and outputting, all at incredibly fast speeds. Because the text is really just sequences of binary numbers behind the scenes, any character, from any script can be assigned to a binary number. Provided the computer is given the map of which binary number is assigned to which character and there is a graphical representation of the character is available (in a font file), the character can be 'word-processed'. In the words of a contemporary computer programmer interviewed for this chapter: 'It's all just arbitrary symbology games.'

Therefore, computers can be multilingual along the lines of mapping characters to binary numbers, ad infinitum, until all characters of all the scripts are mapped, and fonts are created for all those characters. However, there are problems when we closely consider the dynamics of some scripts of non-linear character construction (like Korean, in which characters are compound constructions – many characters going to make one compound character). Further, some languages don't have alphabets like the English alphabet, but they have a mixture of quasi-alphabet (incorporation of radicals into standard characters) and stand-alone ideographs representing words, as in Chinese. The challenge with non-alphabet based languages is that there are an exceedingly large number of individual characters (ideographs) involved and in such situations, the character map that needs to be generated, so the computer can have a unique binary number assigned to each character, becomes unwieldy. English is relatively simple to deal with, because there are only twenty-six letters in the English language alphabet. Therefore all the letters (upper and lower case) and all the numbers, plus additional punctuation marks, can all be safely stored in a 7-bit character code. The '7-bit' part means that the numbers (in which one stores single characters) are 7 memory spaces (bits) long. Seven memory spaces gives you a binary number that is two to the power of seven code points large ($2^7 = 128$ code points):

In a 7-bit binary number (seven spaces for zeros or ones to fill, '0000000') there is plenty of room to fit the 94 different characters outlined above and there are 34 additional spaces left over to make

Figure 3: Breakdown of 7-Bit ASCII character code.

```
26 upper case letters +
26 lower case letters +
10 numeral digits     +
32 punctuation marks  +
==========================
94 different characters
```

up the full compliment of the 7-bit code space. ASCII uses an 8th bit for parity checking – determining whether data has been successfully transmitted, which means we can call ASCII a 1-byte character set (1 byte is equal to 8 bits).

However, a true 8-bit character set should be able to store 256 code points:

Notice that the 257th number in the sequence count (which would be decimal number 256 because zero is a number on computers), would be a binary number that is 9 bits long ('100000000') therefore making it illegal in our imposed limit of 8 bits of space to store our numbers in.

1-byte (or 8-bit) character

Figure 4: Eight bits of space (8^2).

char. count	decimal number	binary number
1	0	00000000
2	1	00000001
3	2	00000010
4	3	00000011
5	4	00000100
6	5	00000101
7	6	00000111
8	7	00001000
9	8	00001001
10	9	00001010
~ ... ~	... ~	... ~
254	253	11111101
255	254	11111110
256	255	11111111

sets are lightweight in terms of memory usage and are well understood in the world of computing. However, when we need to store character sets for languages such as Chinese, which has thousands of characters, we have to move into the realm of 16-bit (or 2-byte) character sets which give us $2^{16} = 65,536$ character spaces to work with. Another particular constraint of computer architectures is that they allocate memory in units of eight, so the next largest step up from a 1-byte character set is a 2-byte character set, even if all of those 65,536 code points are not used up. This is the case in the main Japanese character set today, which is a 2-byte character set that only defines 6,879 characters within the possible 65,536 code points. The real problem with 2-byte character sets is that for every character that is stored on the computer, 2 bytes of memory space are used. Therefore, merely by switching to a 2-byte character set, you have halved your available RAM and disk space. This has serious ramifications if we are to suggest that *every* computer moves to a 2, possibly 4-byte character set, such as the recommendation being made by the Unicode consortium.

At the heart of Unicode is a character code (or character set, as they are called today – and in fact there are 3 levels of the Unicode character set). Unicode, and its related concepts and features, are best understood by tracing the history of computing, from the earliest formalisation of grammars and languages, expressed as a sequence of binary numbers, through to the present difficulties in managing these predefined rules of transformation.

ORIGIN OF CHARACTER CODES:
TELEGRAPH SERVICES AND DATA PROCESSING

Between 1823 and 1838, the American, Samuel Finley Morse (1791–1872) performed experiments with character codes. His work was to enable the encoding and decoding of textual messages sent over telegraphic wires normally, in those days, accompanying train tracks. The mature character code refined from this early research became known as Morse code, heralded by its familiar 'beeps' and 'bips:' dashes and dots.

Figure 5: A Subset of International Morse Code.

A	· —	N	— ·	0	— — — — —
B	— · · ·	O	— — —	1	· — — — —
C	— · — ·	P	· — — ·	2	· · — — —
D	— · ·	Q	— — · —	3	· · · — —
E	·	R	· — ·	4	· · · · —
F	· · — ·	S	· · ·	5	· · · · ·
G	— — ·	T	—	6	— · · · ·
H	· · · ·	U	· · —	7	— — · · ·
I	· ·	V	· · · —	8	— — — · ·
J	· — — —	W	· — —	9	— — — — ·
K	— · —	X	— · · —	,	— — · · — —
L	· — · ·	Y	— · — —	.	· — · — · —
M	— —	Z	— — · ·	?	· · — — · ·

A parallel development in British railway telegraphy, first appearing in 1837, worked toward the same goal as Morse, although it was engineered to require five wires in place of Morse's one. The contraption, belonging to Sir Charles Wheatstone and Sir William Fotergill Cooke, used the five wires to send signals to be received by a deflecting magnetic needle that would move according to a signal and point to a letter of the alphabet on a dial. The 'Morse code' system was simple in comparison, using only one wire to send a signal. The signal would be received by an electromagnet that would attract a small armature in response. In hindsight, a comparison between the American and British solutions shows that the reliability and simplicity of Morse's system was more portable than the relatively sophisticated (hence less-reliable) hardware configuration used by Wheatstone and Cooke. The comparison is also a demonstration of the relationship between software and hardware development cycles, and of the effect both the software and hardware combination have on end-user workflow. Where Morse's system required an operator to translate and transcribe messages from signals to Roman letters and Arabic numerals, the British model automated the translation from signal to letter at the expense of hardware simplicity. The time gained from automating the lookups of signal-to-symbol tables (effectively achieved by printing symbols (letters and numbers) on the dial-face underneath the needle registering signals) in the British model did not warrant the acceptance of complications arising from its failing hardware. Morse code became the preferred model and was refined over time, avoiding significant revision until 1874, and is still in isolated use today.

On May 24, 1844, Morse sent the first message on the first U.S. telegraph link, 'What hath God wrought?' His message explained the popular attitude toward real-time long distance communication in the mid-nineteenth century. However, his aggrandisement of the technology masks both its routine nature, and the more profound formalisation of social language systems. The practice of encoding sequential symbols (language) into a loosely structured flow of dots and dashes on paper gave Morse the ability to exploit the properties of the lightning-fast movement of electrons in an electrical circuit to represent those dots and dashes. He achieved this by prefacing the observance of the circuit's state (flowing or not flowing) over time with a clause that decoded the observed pattern of state-over-time faithfully back into sequential symbols – language. That connection was the spark toward the mass telecommunication networks that we observe (or otherwise interact with) today. During transmission of a *Morse-coded* message the following rules apply:

- a dash is three times the duration of a dot;
- an interval equivalent to a dot separates the individual dots and dashes that make up characters (code point differentiation);
- the start and finish of a sequence of dashes and dots that make up a character are cushioned by an interval of three dots (character differentiation);
- the signal for a space between words is equal to six dots (word differentiation).

When creating the character code, Morse optimised the encoding of the most frequently used letters by assigning them the shortest sequence of dots and dashes. The letter 'E' in Morse code is simply one dot, while a question mark is two dots then two dashes then two dots. Morse based his judgement of the most frequently used characters by counting the individual pieces of type in a printer's draw. Therefore the code is highly customised to suit the needs of English spelling and grammar. Typically, this type of optimisation of binary number symbology to natural language symbology continues to build upon a character code until 'optimum' standard is reached. English language Morse code was customised through several versions (Early, American and International Morse Code standards) just as the Chinese Telegraph Code, which mapped 9800 Han ideographs (Chinese characters), went through significant changes from its first appearance in 1871 to the end of the electrical telegraph era. The existing International Morse Code cannot be bettered without the conceptual leap that occurred during its optimisation process. Because Morse code requires a message to be encoded by a person in real-time at one end of the wire, and decoded in real-time at the other end of the wire by a second person concurrently writing the message on paper, automation of encoding and decoding was desirable. This automation was achieved by recording messages on a telegraphic typewriter that would make representations of the input from the keyboard as perforations in paper tape. The perforated tape was then fed into a machine that would read the perforations and send the appropriate Morse code sequence, which would be subsequently picked up by a second machine at the receiving end, and written back as perforations in a duplicate tape. Perforated tape reading and writing machines made possible transfer speeds of up to 400 words per minute by 1900.

The world's first binary character code was pre-empted by Morse's code for numbers. While Morse used a frequency algorithm to apportion codes of varying bit lengths to Roman characters, each of the Arabic number characters had identical frequencies of use, as such Morse gave each number a code containing the same number of bits, as there was no advantage in making some numbers faster to encode than others.

Figure 6: The 5-bit Morse code numeral characters.

Jean-Maurice-Émile Baudot applied the idea of a 'fixed-length' character code for all characters in 1874 creating the Baudot Code in France. The Baudot Code was developed for use with the then 'next-generation' telegraph device, the teleprinter (which Baudot also designed). Early teleprinters used a five-key keypad to encode characters on narrow two-channel transmission tapes. Baudot's character code displayed the uniformity and 'control' sequences that contemporary computer character codes use and also employed a shift-locking device that gave each binary number the ability to represent a letter and a figure.

QWERTY-style keyboards were adapted to automatically punch the correct 5-bit binary sequence into the transmission tape, making the text encoding process much faster. Further, the Baudot system made efficient use of the telegraph line by 'multiplexing' or sending up to six individual messages over the one wire at the same time by using a time-sharing device. Baudot's work became the cornerstone of telegraphy until the entire textual telegraph was made temporarily redundant by Graham Bell's famous invention of 1876 – the telephone system (which was the result of work into improving the textual telegraph). While the telephone is still very much an obvious part of our lives today, the textual telegraph ceded valuable encoding practices to the con-

Figure 7: A 5-bit Baudot Code Set.

Binary Value	Letter	(Shift) Figure
00011	A	–
11001	B	?
01110	C	:
01001	D	$
00001	E	3
01101	F	!
11010	G	&
10100	H	STOP
00110	I	8
01011	J	`
01111	K	(
10010	L	)
11100	M	.
01100	N	,
11000	O	9
10110	P	0
10111	Q	1
01010	R	4
00101	S	BELL
10000	T	5
00111	U	7
11110	V	;
10011	W	2
11101	X	/
10101	Y	6
10001	Z	`
00000	N/A	N/A
01000	CR	CR
00010	LF	LF
00100	SP	SP
11111	LTRS	LTRS
11011	FIGS	FIGS

struction of the modern computers, even Baudot's name is enshrined in the term we use to designate the number of data signalling events that occur in a second – the Baud-rate.

So it was in another area of technology development, outside telegraphy, in which the character code found its most beneficial use. In 1880, the United States Census Bureau contracted Herman Hollerith (1860–1929) to work as a statistician. While working with the Census Bureau, Hollerith devised the Hollerith Code, a character code that was used to encode alpha-numeric sequences of census data on the punch cards that were used as memory by his early tabulating machines. Though the reading and processing of the punch card by the tabulating machine was cumbersome, remarkable speed increases were gained over the manual calculations nevertheless. By contrast, the 1880 United States Census that was hand calculated, took seven years to produce, while the 1890 United States Census, that employed Hollerith's tabulating machine using the Hollerith Code, took only six weeks to produce even more detailed information. With such a dramatic demonstration of the power of mathematically processing the census records – the United States Census Bureau posted a US$5 million saving in 1890 attributed to the automation – Hollerith capitalised on his invention and founded a company called the Tabulating Machine Co. in 1896.

In 1911, Hollerith's company merged with the Computing Scale Co. and the International Time Recording Co. to form the Computing-Tabulating-Recording (CTR) Co., and over the course of the following seventeen years, forged new markets for data processing on several continents. The expansion of CTR Co. into transnationals enterprise was triggered by Hollerith's Roman character code that sufficed for the majority of Latin-based languages without major modification.

CTR Co.'s ability to significantly improve international business practice, and international businesses' performance, beyond government and military, lent support for a name change. International Business Machines (IBM) Corporation was decided upon in 1924, just five years before Hollerith's death, and remained a landmark to computing for the next fifty years. Hollerith's character code and punch card system, being the backbone to most of IBM's technologies, enjoyed continued use until the character code underwent major revision and localisation to fifty-seven different regions starting in 1964. The code, which was re-released under the name Extended Binary Coded Decimal Interchange Code (EBCDIC) effectively isolated itself from the market by ignoring the advice of the standards promoting bodies who foresaw the benefit of, and need for, a common and international platform for text exchange.

CHARACTER CODES FORMALIZED

ASCII, ISO 646 and EBCDIC

The American Standards Association, later to become the American National Standards Institute (ANSI), first conducted research into character codes in the late 1950s. Their initial ambition was to capture the range of characters that were presented on the keyboard of an English language typewriter. A 7-bit character code was devised, offering 128 code spaces (number of unique binary numbers possibly stored in 1 to 7 orders of magnitude) to store upper and lower case characters, numbers and punctuation marks. An incomplete (no lower case) character mapping was published in 1963 as the 'American Standard Code for Information Interchange' (ASCII). ASCII was refined over the next five years mapping 32 control characters and 96 printing characters, which is a reflection of its present mature state bar the later modification of ASCII to 8-bit, accommodating 190 printing characters. IBM was the only computer manufacturer from the United States to *not* adopt the American Standard Code for Information Interchange. Being the stalwart of the computer industry, IBM had legacy mainframe computers using various revisions of the original Hollerith code that they were obligated to support, and they had a large enough hand in the market to exclusively support their own new character mapping mentioned above, EBCDIC. However, this behaviour was seen as 'proprietary' and frustrating to the end users, who were disadvantaged by the introduction of IBM's updated character code, not being able to simply share data from IBM computers with other computers and *vice versa*.

Computer manufacturers in the United States during the 1960s and 1970s were the world's most advanced technologists and the world's most successful exporters of their equipment. Through the export routes of all United States computer manufacturers (except for IBM), ASCII and the QWERTY keyboard were disseminated into the outside world, slowly cementing themselves as the *lingua franca* of global text processing and input. The prevailing understanding of ASCII saw use in making further minor modifications to include the characters needed for Western European localisation and more general interlingual Latin alphabet support. By 1967, the Swedish body, International Organization for Standardisation (ISO), had constructed a simple multilingual schema to reserve ten spaces in standard ASCII for national variants. ISO released this framework as ISO Recommendation 646, and it allowed ASCII to be

Figure 8: 96 printing characters of 7-bit ASCII.

```
==============
|_|0|@|P|`|p|
|!|1|A|Q|a|q|
|'|2|B|R|b|r|
|#|3|C|S|c|s|
|$|4|D|T|d|t|
|%|5|E|U|e|u|
|&|6|F|V|f|v|
|'|7|G|W|g|w|
|(|8|H|X|h|x|
|)|9|I|Y|i|y|
|*|:|J|Z|j|z|
|+|;|K|[|k|{|
|,|<|L|\|l|||
|-|=|M|]|m|}|
|.|>|N|^|n|~|
|/|?|O|_|o| |
==============
```

used as the basis of character codes that required different scripts, such as Greek and Arabic. In 1969, ASCII was incorporated as the basis of Japan's major character encoding scheme, JIS. To date there have been over 180 language variations of ASCII.

Although ASCII has become the *de facto* character-encoding standard, it generally operates as the basis of a character set rather than its definition. Microsoft and Apple have both produced their own variations on 7-bit ASCII that are enlarged to an 8-bit (1 byte) code space and include special characters to that allow their character sets to be used seamlessly for the input of other Latin-based scripts such as Italian, Spanish, French, Swedish, etc. However, such synecdoche has limitations: one character code to several languages, all crammed into the 8-bit code space doesn't allow full support of localisation characters like currency symbols, language-specialised typographical ligatures and so on. Therefore, while Microsoft and Apple can support most Latin-based languages (characters like é, å, ß etc.) with their standard character sets, it is a piecemeal approach and forces clumsy typographic 'tricks' to assail shortcomings (for example, using inline, hand crafted bitmap images to stand-in for missing characters).

SUPPORT BEYOND ENGLISH
(ASCII TO ISO 646 TO LATIN-1)

The various accent and other diacritical marks that are not supported in the two main, ASCII based, Microsoft and Apple character encoding are undeniably necessary for the accurate digitisation of Latin-based scripts (other than English). Moreover, languages that are not based on Latin script, like Greek, Hebrew, Cyrillic, Arabic etc. are completely unsupported under ASCII and its extensions through ISO 646. Computer manufacturers and the International Organization for Standardization undertook further work to produce a methodology for extending 7 and 8-bit ASCII-based character codes to encompass the hitherto unsupported characters of non-Latin scripts (including ligatures, currency marks, etc.). The results of that work were formalised into a recommendation named ISO 2022 that defined how to build character sets (and in reverse, from the computer manufacturers perspective – *how to support character codes*). ISO 2022 was used to extend ISO 646 into many different character codes, one of the most popular, ISO 8859-1, is still in use today. ISO 8859-1 is affectionately known as 'Latin-1,' and is the *de facto* standard for data communications in Western Europe. 'Latin-1' is only the first part of ISO 8859, a fifteen-part (and growing) character code. Each part relates to each other as though they were different 'type heads' on the one typewriter that can be seamlessly interchanged maintaining core functionality while replacing the representational imprint from the same input channel (key on a keyboard). Because each character set is like a 'type head' of a typewriter, they can be used independently of each other, but it is difficult to use them in combination. The fifteen parts of ISO 8859 are as follows:

Part 1:	*Latin Alphabet No. 1* (Revised 1998) Western European Languages (Latin-1)
Part 2:	*Latin Alphabet No. 2* Eastern European Languages (Slavic, Romanian, Albanian, Hungarian
Part 3:	*Latin Alphabet No. 3* Southern European Languages (Maltese) and Esperanto.
Part 4:	*Latin Alphabet No. 4* (1998) Northern European Languages
Part 5:	*Latin/Cyrillic Alphabet*
Part 6:	*Latin/Arabic Alphabet*
Part 7:	*Latin/Greek Alphabet*
Part 8:	*Latin/Hebrew Alphabet*
Part 9:	*Latin Alphabet No. 5* Latin character set used for modern Turkish
Part 10:	*Latin Alphabet No. 6* (1988) Icelandic, Nordic and Baltic character sets
Part 11:	*Latin/Thai*
Part 12:	*Reserved for Latin/Indian*
Part 13:	*Latin Alphabet 7* Baltic Rim, Latvian and local quotation marks.
Part 14:	*Latin Alphabet 8* Gaelic and Welsh – with Latin-1 covers all Celtic languages.
Part 15:	*Latin Alphabet 9* Latin9, nicknamed Latin0, updates Latin1 by replacing the symbols ∫, ¨, ´, 1/4, 1/2 and 3/4 with forgotten French and Finnish letters and placing the Euro currency sign in place of the former international currency sign.

It is crucial that there are standards (as above) that define how computer manufacturers, computer programmers and computer users encode and decode their data – otherwise computers (or systems in general) cannot interoperate, nor can the one hardware be used to support a series of language specific character codes (for example, the different *parts* of ISO 8859). This is indeed a topical issue, one that is increasingly looking for some kind of *enforcement*.

The World Trade Organization (WTO) is the international organization dealing with the global rules of trade between nations. Its main function is to ensure that trade flows as smoothly, predictably and freely as possible.

ISO – together with IEC (International Electrotechnical Commission) and ITU (International Telecommunication Union) has built a strategic partnership with the WTO. The political agreements reached within the framework of the WTO require underpinning by technical agreements. ISO, IEC and ITU, as the three principal organizations in international standardisation, have the complementary scopes, the framework, the expertise and the experience to provide this technical support for the growth of the global market.[1]

Generally speaking, there are no bureaucratic bodies (WTO or otherwise) with a scope and mandate to provide for such enforcement of character code use and compliance. Computer manufacturers and software developers have always done exactly as they wished, taking from the standards some useful guides, and extending the standards into proprietary character encoding structures that suit whatever market demands need to be addressed at the time. The widespread adoption and growth of standards therefore, do not accord to specific technologies or encoding technology; they accord to the general technologies that allow for flexibility of use and transport. Hence, standards in today's computing environment are always *de facto*, they are always an electronic equivalent to a natural language's *vernacular*. An example is the hypertext transfer protocol (http) that negotiates the transfer of data over the Internet. The first thing that happens when a page is requested by a browser from the Internet is a 'handshake' between the client and browser that mentions, among other things, what the character encoding of the data about to be exchanged will be. This means that that http standard can be effectively implemented everywhere without fear of data deprecation. Moreover, the American computer industry *is* the global

---

[1] http://www.iso.ch/

computer industry, so computer users everywhere must accept and comply with Apple's and Microsoft's truly dominant hardware and operating system software requirements – ISO and WTO (not to mention IEC and ITU) are merely a skirt to the protagonists.

IDEOGRAPHIC CHARACTER CODES
(CHINESE – JAPANESE – KOREAN)

The QWERTY keyboard has come into widespread use all over the world. It is based on the modern Latin alphabet, and it obviously does not directly support the input of the tens of thousands of ideographs from Chinese, Japanese and Korean languages. The question might be asked then, why are ideographic representations desired in a computer environment? Ideographs pose a difficult input problem (in terms of the limited keys on a standard keyboard) and there are mature romanisations of these ideographic languages that could make use of existing word processing software and hardware. The problem with romanisation systems (and the same applies to speech) is that they do not effectively transliterate homonyms, in that the meaning is only seen in the ideograph, not in the phoneme. Therefore, the romanisation systems that are used, like Pin-Yin (pin meaning 'spell' and yin meaning 'sound') suffer from the same problems that speech does, only that in speech these things are negotiated as humour, whereas incomprehensibility in written language is seen as useless. Nonetheless, Pin-Yin spelling is used extensively on signs and posters throughout China although the accent marks that are needed to reproduce the tones of the language are seldom used. Almost all Western newspapers, from the *New York Times* to *The Australian*, have adopted the Pin-Yin spelling to render Chinese names and terms. It is remarkable that such enormous publishing regimes remain technically and editorially incompetent at rendering and proofing scripts that are dramatically different to those addressed in, for example, the parts of ISO 8859 and it is further testimony to the substantial difficulties that are posed by the typesetting and handling of ideographic texts. However, the more specific problem we see at work in the newspaper scenario is the need for, but impossibility of, mixing different character codes within the one document. The answer is to mix characters from different languages in the one character code, as opposed to mixing different language character codes in the one document.

Like all character codes, East Asian language character codes have their origin in the early telegraph technology. Although romanisation systems were well established at the time, it was more cost effective to develop character codes that directly mapped the main ideographs to Morse-like codes (and these often entailed encoding more than ten thousand individual characters).

Modern Japanese character codes were developed by the Japan Industrial Standards Committee (JISC) from the 1970s onward. The first Japanese language character code was an extension of ISO 646 that supported *katakana*, just one of the three Japanese syllabaries (which are *hiragana, katakana* and *kanji* [the Chinese characters used in Japanese]). This code was refined over several years to a stable release known as JIS X 0201-1976. The logical extension to this initial code was to include *hiragana* and some of the *kanji* characters. Over the next fourteen years the JIS X character codes grew to maturity, including *hiragana, katakana,* and several levels of *kanji,* and also Greek, Cyrillic and some Eastern European characters beyond the scope of the foundational ISO 646 characters. Therefore work on the early JIS X character sets was actually the first work toward a *multiscript* character set that would be the solution to the newspaper problem outlined above.

The definition of *levels* of *kanji,* mentioned above, defines as Level 1 a limited set of ideographic characters for 'everyday use' to simplify computer memory management of the character code. Further *levels* define more obscure ideographs, but they are implemented as separate character sets, so the one is not accessible from the other. As can be appreciated, this creates an interesting retardation of the Japanese *kanji* vocabulary for the benefit of computer performance. It is suggested that character encoding techniques and practice shape the possible use of a language (and as such, society's image of itself) in quite profound ways, more obvious in the implementation of gargantuan character sets that need to be reduced to 'everyday use'.

Further complications with JIS X *multiscript* character sets are seen in the side-by-side use of 7-bit, 8-bit and 16-bit characters within the one encoding. As originally specified by JIS, the different bit length characters were to be distinguished from one another by 'escape sequences' within the text. As it is difficult to continually interrupt the flow of text input with banal demarcations between the 7-bit ISO 646 characters, 8-bit *katakana* and *hiragana* and 16-bit *kanji* characters, Microsoft designed a 'workaround' that removed the use of escape sequences, but retained the original character mapping. This new system of encoding the JIS X character set was called Shift-JIS and it effectively encapsulated the escape sequences inside the encoding itself by defining binary distinct value ranges for single byte character and double-byte characters, so when a certain character was input (via a range of different methods – direct keystrokes, multiplier keys, combination strokes and so on) the computer would be able to judge whether it was a single byte or double byte character that it had encountered. While Microsoft had provided a more streamlined text input workflow with Shift-JIS, they had also severely limited the number of *kanji* characters that could be encoded in the set (again, deforming language patterns and expression, let alone translation issues). Moreover, the Shift-JIS only defined value ranges to designate 8-bit and 16-bit characters. 7-bit characters would not be able to be understood by a system supporting Shift-JIS, making it entirely unusable for data transfer in an environment like the Internet, where there are many exotic character codes being used: 7-bit or 8-bit or 16-bit or even 32-bit codes spaces, and for any one of them to not be supported would mean loss of data through corruption.

More practical variations of JIS X were developed for use in client-server environments, one of which is known as ISO 2022-JP and widely used for email and textual data transfer. Another is the 8-bit Extended Unix Code-JP (EUC-JP) in which a wrap-around scheme was devised to allow East Asian scripts to be modelled around conventional versions of the ASCII character set by *adding values* to each of the East Asian language character sets so as to distinguish them from ASCII forms. Extended Unix Code can in this way be used to map many scripts side-by-side without the use of escape sequences and without the danger of confusion between different bit length characters.

The difficulties that were encountered in the early days of Japanese text processing highlighted a path of least resistance for the development of Chinese, Korean and Taiwanese character sets. The People's Republic of China uses in its national script the largest number of *simplified* Han ideographs out of any of the East Asian languages. The *simplified* ideographs are borne out of the language-reforms of the 1950s aimed at increasing national literacy. As such, the character set for the People's Republic of China has come to be known as 'Simplified Chinese'. Taiwan, on the other hand, is still using the original form of the Chinese ideographs for its national script, and hence has its character code dubbed 'Traditional Chinese'.

While Simplified Chinese and Traditional Chinese character sets were able to learn from the development of the Japanese character sets, the main Korean language script, *hangul*, brought new challenges to the construction of character codes. *Hangul* is an alphabetised script, although out of its alphabet it builds not so much words, but syllabic blocks of stacked characters that look similar to the fixed ideography of the Chinese scripts. So there are two possibilities, one is to create a system through which the syllabic blocks are hand crafted by typing each letter component, the other is to create a character set that has all the possible combinations of the letters prearranged and key the data through a traditional QUERTY keyboard using Romanisation systems (in the same way Chinese ideographs can be entered into a computer). The former process has prevailed and the main Korean character set includes a massive collection of precompiled syllabic blocks.

TOWARDS UNICODE

So far we have seen progressive phases in the development of text processing, from telegraphic services to ASCII and its formalisation through ISO, to national variants and extensions to multiple encoding strategies to the point where we can draw a conceptual map of the world's use of 'computable' representations of scripts. The possibility of such a mental picture of the world's authors and recordists capturing textual data in the various ways that they do, producing texts containing many different scripts, coincides with another mental map, that of the interconnectedness of all of those text processing machines (computers) via the Internet. While the input and capture of data remains highly localised, the *reach* of that data is global. It is imperative, for that reason, that we acknowledge in the preparation of our highly localised textual content the imminent possibility that data will be collected in repositories that have no regard for our data's particular parochialism, and that our own processing structures will most likely, sooner or later, play host to other foreign national data.

The need for a forum on standardization – part think-tank, part database – was identified by the immediate past President of ISO, Prof. Giacomo Elias, that would bring together the UN agencies, NGO's and private sector – all those, in fact, developing, working with or using standards. The idea was to streamline work between the bodies involved, to avoid duplication of effort, to harmonize standardization projects, and plan ahead rationally together. To find two agencies working independently of each other on similar projects, each unaware of the other's activity, is wasteful of resources and time.[2]

Just as IBM's EBCDIC character code was shunned by users for its lack of compatibility with other systems, Microsoft and Apple are now finding that their own hand-rolled character sets are suffering from similar user-antagonism. Proprietary data encoding and formats are made more apparently problematic in a networked environment that connects computers of all kinds.

In response to several factors: user-antagonism, a networked-society and a growing need and possibility of sharing data between business, government and other organizations (in which are included private individuals) the development of a universally adoptable, multilingual, multipurpose character code has become an obvious and unavoidable process. Taking a lead from the early work on multilingual character codes undertaken by Xerox in the early 1980s for their Star Workstation, United States computer software and hardware firms gathered enough combined energy to launch an Industry wide initiative called Unicode to create a super character code of all the world's existing character codes, and refine them as necessary. Unicode by itself has never tried to specify how text systems should operate, only determining what the character code breaks down as, and how software engineers should aim to use the code.

The dominant computer industry bodies, predominantly in the United States, have worked for nearly twenty years in think-tanks to create and expand the use of universal mappings for more than just how characters should be encoded (which is the domain of Unicode), but also how a sequence of characters should be composed into a file and how the file's structure and metadata should be tagged (a 'high level' semantic process) so as to provide a natural-language and software-vendor neutral networking environment.

---

[2] http://www.iso.ch

CASE STUDY: ABOUT UNICODE

Unicode provides a unique number for every character, no matter what the platform, no matter what the program, no matter what the language.

Fundamentally, computers just deal with numbers. They store letters and other characters by assigning a number for each one. Before Unicode was invented, there were hundreds of different encoding systems for assigning these numbers. No single encoding could contain enough characters: for example, the European Union alone requires several different encodings to cover all its languages. Even for a single language like English no single encoding was adequate for all the letters, punctuation, and technical symbols in common use.

These encoding systems also conflict with one another. That is, two encodings can use the same number for two different characters, or use different numbers for the same character. Any given computer (especially servers) needs to support many different encodings; yet whenever data is passed between different encodings or platforms, that data always runs the risk of corruption.

The Unicode Standard is a character coding system designed to support the worldwide interchange, processing, and display of the written texts of the diverse languages of the modern world. In addition, it supports classical and historical texts of many written languages.

Formally, the Unicode standard is defined by the latest printed version of the book The Unicode Standard, plus online documents and data that update and extend the book's normative specifications and informative content. The latest version of the Unicode Standard is Version 3.1. The primary feature of Unicode 3.1 is the addition of 44,946 new encoded characters. Together with the 49,194 already existing characters in Unicode 3.0, that comes to a grand total of 94,140 encoded characters in Unicode 3.1.

The new characters cover several historic scripts, several sets of symbols, and a very large collection of additional CJK ideographs. Unicode 3.1 also features new Unicode character properties, and assignments of property values for the much-expanded repertoire of characters.

Source: www.unicode.org

CONCLUSION

As a global market we are still in an embryonic stage of development in terms of our widespread understanding of how to use technology, what to use it for and what the cultural, social and economic implications are. While all the latest release, major computing platforms support the Unicode character set and even provide fairly mature applications and fonts to manipulate, access and display the characters themselves, its eventual strength in distributed publishing environments is yet to be tapped. By combining the key concepts elucidated in this chapter and the previous chapter: those of a universal character set and the separation of *presentation infor-*

mation and *content information* of a given set of data, we are provided with the beginning of a truly open, borderless, stateless, cultureless – in short neutral – publishing environment. Only with such a neutral environment can a properly tooled networked society flourish.

Chapter 7

MANUSCRIPT USE AND TYPESETTING ISSUES

Atsushi Takagi

The globalisation of digital technologies creates the necessity of multilingual technologies. This chapter discusses some of the practical issues arising in a multilingual typesetting environment where the core technologies of computerisation have been created in the English-speaking world.

Numerous countries in Asia have seen the importance of working with digital technologies and building the necessary infrastructure and skills base of the new economy. Japan's IT strategy headquarters has declared an 'e-Japan strategy' the aim of which is to ensure that 'Japan becomes the most advanced IT nation in the world within 5 years'. Japan has taken the first step to promote this IT revolution with the 'e-Japan emphasis plan', to be followed by a concrete action plan in March 2002. In Korea, nearly 50% of online Korean households are connected to broadband Internet services. Korea is the world's leader in broadband access with over 6.25 million subscribers. Singapore's Prime Minister, Goh Chok Tong, has also promoted entrepreneurship and the new technologies, using the US$1 billion Technopreneurship Investment Fund. Since Singapore is a smaller country, CATV and optic cables have already been set up everywhere and are run by highly competitive IT industries.

As at mid 2001, China has invested US$2.5 billion to develop its broadband infrastructure. The stretch of fibre optic cable is expected to reach 16 million kilometres over the next five years. According to a senior Lucent Technologies executive, China is expected to invest $100 billion over the next five years on upgrading all aspects of its telecommunications systems as increased market deregulation creates massive demand. Malaysia has created the 'Multimedia Super Corridor' to reach new technology frontiers, partnering global IT players. The new concept of electronic government, Cyberjaya, was also introduced – an intelligent city with multimedia industries, Research and Development centres, a Multimedia University, and operational headquarters for multinationals to direct their worldwide manufacturing and trading activities using multimedia technology.

THE SIGNIFICANCE OF TEXT GENERATION TECHNOLOGIES

Considering such rapid developments in IT industries in the Asia Oceania region, any global and local businesses involving IT need to provide information in the vernacular in order to create a universal means of communication. This requires multilingual typesetting. Internet Names WorldWide (INWW) General Manager, Clive Flory, points out that: 'Obviously there are many companies and individuals who speak Chinese and Japanese, so the potential is huge'. Acceptance of the reality that not only English is used worldwide leads quickly to the conclusion that other languages are essential in Web construction and publishing if one is to expand one's market globally.

Multi-language character software such as Global IME (non-ASCII characters: Input Method Editor) and the international character sets such as Unicode, and are now widely used for non-Roman scripts. It is even expected that Internet domain names will soon become available in non-Roman scripts.

As one of many Japanese residents in Australia who use personal computers in their home and offices, typing Japanese characters has always proved to be problematic. I brought a desktop computer from Japan to Australia (and paid heavy import duties) a decade ago. Fortunately, the Japanese community in Melbourne provided me with tremendous assistance for setting up this computer. We decided that WindowsJ ('J' denoting Japanese version) and MacOS (English) with Kanji Talk were the most reliable operating system. There is a huge demand for computer services and increased use of the Internet in parts of Australia with sizeable Japanese populations, such as Sydney, Melbourne and Brisbane. Many computer shops dealing with Japanese clients have flourished. Other communities, such as Chinese and Korean, have similar computer networks assisting with their typesetting and computer related needs.

Chinese, Japanese, Korean and Vietnamese texts, (named CJKV by Ken Lunde, 1999), are displayed by more than a single byte (more than 8 bits). These text-input systems are still in the development stage and CJKV encoding and decoding systems are not yet universally recognised. Unicode is now considered the most stable 16-bit character set for those tasks, and is used in Windows 98, ME and 2001.

Written Japanese consists of four types of characters:

- Hiragana
- Katakana
- Kanji, or Chinese characters
- Roman characters

The use of these four sets of characters can be considered a unique orthographic system, compared with say, Chinese and Korean writing, which use fewer than 3 syllabaries. To type Japanese characters, one needs to use Roman keyboard to type in the word phonetically, for the word 'watashi' (meaning 'I' in English), this involves typing the Roman letters w-a-t-a-s-h-i then pressing the space bar to convert the Roman letters into the Japanese alphabet of your choice.

There are two ways to have a CJKV operable system. One is to install CJKV support software on your English computer; and the other is to replace your entire operating system with a non-English version. The more stable way to practice CJKV typesetting is to obtain a local operating system that provides CJKV assistance. In the case of MacOS-J (formerly Kanji Talk), there are few difficulties, since it provides a complete Japanese environment. Non-English versions of Windows operating systems include localised menus and dialog boxes, and also TrueType fonts. Windows 2000 has a Unicode function, allowing the user to type and read Japanese, Chinese, Korean and Vietnamese, etc. The only condition is that those 2 byte fonts are used only on Microsoft software. Adobe and Macromedia also support 2 byte fonts.

The web is serviced by various shareware and freeware, enabling characters to be entered into and read on websites such as AsiaSurf, NJStar, NJWin, UnionWay, or Global IME for Windows 95 and 98. In order to use Japanese and languages other than English on the web, the user depends on browsers that support multilingual text, such as recent versions of Internet Explorer. Through this browser, the user can display and enter Japanese characters, provided Microsoft Internet Explorer is installed (4.0 or a higher version) and Microsoft Global IME is on the computer's Windows 95, 98 or NT. Further information regarding double byte font setting on any other operating systems, such as LINUX, UNIX TRON can be found at

the Ken Lunde (CJKV information processing) site
<http://www.oreilly.com/people/authors/lunde/>.

The remainder of this chapter consists of case studies highlighting some of the critical issues in multilingual typesetting. While the case studies mainly involve the Japanese language, the issues emerging are representative of those that will be encountered by anyone dealing with non-Roman/non-ASCII text.

CASE STUDY 1: OPAL JAPANESE LANGUAGE CD PROJECT

In 1995, the National Asian Languages/Studies in Australian Schools (NALSAS) taskforce was established in order to develop a flexible delivery program of teacher training in the Japanese language. The project commenced in January 1996 and ended in October 1998. During this time, course developers and Universities produced Japanese language course materials at three levels:

- Access Program (160 hours)
- Graduate Certificate (200 hours)
- Graduate Diploma (200 hours)

The Japanese language materials were developed to support these programs, and were designed to be of particular relevance to classroom teachers of Japanese, using a communicative methodology for adult language learners and combining new technologies for the distance teachers learning. The project was called *OPAL: Japanese.*

The material was developed by communicating via email, teleconference, and face-to-face meetings between three Universities with established Japanese language Departments. Since the *OPAL: Japanese* courses were designed for distance delivery to students anywhere in Australia, the project team had to create the following course materials and delivery modes:

- Course textbook including workbooks
- CD including video, visuals & script and audio
- Audio tapes

Students use email, phone and fax for teacher-student or student-student consultations, and assessment is done on the content of the workbooks.

A number of major issues emerged during the development of the material. Communication required adequate email software, and setting the Japanese fonts on word processing software, CD and textbook materials in order to successfully transfer files backward and forward between the team members. Some of the worst communication, we discovered, takes place between computers. It was a frustrating experience to receive documents in a format one's computer cannot open nor read. It was crucial to determine hardware, software, and other platform-related standards prior to the project.

First, our team members discussed the best way to communicate using email, which was a reasonably cost effective and instant means of transferring team members' messages across the country. This meant we had to standardise our email software. Although we concurred on the type of browser and email software, team members still had to learn how to use it, and the process was not without its difficulties. On several occasions we received Mojibake (unreadable text) email or files, but eventually rectified this by using the common encoding methods.

In order to send and receive the created documents, we concluded that the members would use WordJ (Japanese version of Word 6) on Windows 95 operating system, and Kotoeri and Kanji Talk software installed on MacOS using English Microsoft Word 6. We also used MacOS which had a system partitioned into English and Japanese operating systems. When we were writing materials on Word documents we had to follow a number of agreed protocols, as we had to aim for consistency in styles of formatting, fonts, font sizes, tabs, etc.

We found that although we agreed to use the same Microsoft Word software, we still had some typeface style mismatching between Mac and PC. On the basis of this condition, we started communicating with other members of the project in both English and Japanese. To avoid any mis-transaction of Japanese fonts in a mix of Japanese and English texts over email (and as we were creating language learning materials, most of the text being transmitted was multilingual), the team members sent document files as text or rich text format (rtf) attached to the email.

The desktop publisher who produced the final text insisted they could use QuarkXpress. However, since it was an English version of software, it had difficulty in establishing Japanese typesetting. There are some brochures and newspapers produced in Australia for Japanese tourists that are done by QuarkJ (J = Japanese) on a Japanese OS (WindowsJ). However, in the OPAL project we finally settled on running WordJ on a Mac with a Japanese operating system called KanjiTalk. This is how we prepared and published the print material. Word was not particularly flexible when it came to inserting images, but we achieved a satisfactory result.

CASE STUDY 2: LONELY PLANET JAPANESE GUIDE BOOK AND PHRASE BOOK

In 1997, I was involved as a translator with the major city guide publishing company, Lonely Planet, which was at the time updating both its Japan guidebook and Japanese phrase book. In this process, a number of issues arose regarding software compatibility and the transference of files between a proofreader and a manager of guide and phrase book. The proofreader was advised to use International Language Kit software, enabling him to set up Japanese (or languages other than English) typefaces on an English OS, and this was tested. However, the proofreader had Japanese OS installed on his computer and so typed Japanese characters which were eventually recognised on Lonely Planet's English OS using the International Language Kit.

Lonely Planet received an electronic document from a Japanese correspondent in Tokyo and the proofreader checked all the names and phrases on a hard copy. While transferring the document by floppy or email attachment, a problem was discovered: the newly-edited document saved to disk was somehow converted to unreadable characters and misspelled words. Eventually, we had to re-type the documents using International Language Kit at the Lonely Planet premises. The eventual solution for Japanese typeface input was to use Windows 2000 (English version).

CASE STUDY 3: BLACKWELL SCIENCE ASIA

Blackwell Science Asia is based in Melbourne, Australia, and publishes scientific, technical and medical (STM) print and electronic books and journals for Asian markets in English and Asian languages. Blackwell publications are available in both print and electronic form, and their editorial services cover all areas of medicine in both English and Japanese. Blackwell Science Asia has over the past decade expanded its journal publishing service, publishing a rapidly growing list of national and international journals for the Asia Pacific.

Blackwell Science has recognised the importance of positioning its reputation in the Asian region, opening branches in the region – Blackwell Science Japan was established in central Tokyo in April 1997. There is a demand for Japanese and other Asian languages in medical science publishing, and so Blackwell's Tokyo office was established mainly for the purpose of public relations and liaison with medical institutions and universities.

Late in 1995, Blackwell Science Asia purchased a Japanese NEC computer in order to open and read Japanese and English documents on floppy disks. At this time, NEC was a very common personal computer, pre-installed with Windows 3.1 or Windows 95, and was widely available among medical institutions in Japan. Blackwell Science considered this model would be a more efficient way to receive documents written by Japanese scholars and doctors. The documents were usually saved on NEC formatted floppy disks. Blackwell Science currently practices Mac and Personal Computer software applications, such as Word and Adobe's PDF (Portable Document Format) for exchanging documentation across countries. Almost all documents are published in Hong Kong and Singapore instead of the Melbourne office, which is now mainly concerned with editing.

MEETING THE CHALLENGE OF MULTILINGUAL TYPESETTING

These case studies illustrate a number of the practical issues that arise in meeting the challenge of multilingual typesetting. Most programs and operating systems for computerised text processing have not been created in an English-speaking environment. Multilingual text encoding can occur within an English text-processing framework, or inside an operating system dedicated to a particular language and script. Either way, numerous practical difficulties emerge. Recent technologies have improved this situation, such as the Unicode character set and the more recent versions of Internet browsers, operating systems, word processing programs and desktop publishing programs. These technologies are progressively improving the capacity of computers, and human information and communication systems, to work effectively in the emerging multilingual global information economy.

REFERENCE

Lunde, K. (1999). CJKV Information Processing. O'Reilly & Associates, Inc.

Chapter 8

TRANSLATION IN A DIGITAL ENVIRONMENT

Laurie Gerber

In this chapter, we look at how texts get to be translated, and how this is likely to change in the future. (In this context, 'text' will be defined as any piece of written language.) Behind every text that is available in translation, there is a chain of decisions and processes. How is it that texts come to be candidates for translation? What are the contexts in which this happens? Assuming that the source language (the language in which the text was originally written) is given, how are *target* languages selected (the language(s) the text will be translated into)? The questions here could apply to translation at any point in history, but changes in technology and in the market for translation will change the answers. In short, we can expect the following changes:

- infrastructure developed to facilitate globalisation will simplify translation projects in all areas;
- translation as a profession will enjoy higher prestige, and skilled translators will enjoy better pay, while the types of translation done by humans will shrink;
- language modelling techniques being developed in computer science and computational linguistics will make machine translation a viable option for many more types of translation;
- machine learning of translation rules will make real-time translation available for most texts between any written languages;
- semantic annotation and analysis may alleviate the N^2 problem of language combinations.

ECONOMIC AND SOCIAL FACTORS THAT DRIVE TRANSLATION

Characterizations of the Uses of Translation

In the popular business book, *What They Don't Teach You at Harvard Business School* (McCormack 1985), author Mark McCormack tells an anecdote about meeting the president of Rolex. 'How's the watch business?' he asks. 'I wouldn't know', says the president of Rolex, 'I'm not in the watch business, I'm in the luxury business.' The anecdote led me to think more seriously about the translation business, and what translation means to its consumers. Sometimes translation gives *access* to information. Sometimes translation *enables* business expansion. Finally, translation can make *communication* possible. Along these lines, the translation market can be divided into 'assimilation', 'dissemination', and 'communication' uses of translation.

For 'assimilation', translation is *access*. Access to information that otherwise would not be available. For example, researchers and engineers need access to published developments and research results from the international community in their field. When applying for patents, it is necessary to search international patent databases to anticipate any conflicts that may arise with the patented technology. Governments and their intelligence agencies use translation to monitor the foreign press, and in some cases to spy on the activities of other countries.

For 'dissemination', translation *enables* the extension of commerce beyond borders. The most direct form is the translation of books with international market appeal. But more frequently, translation for dissemination involves product literature such as user and repair manuals that allow the development of export markets for other kinds of products, such as machinery and software.

The use of translation for communication has certainly been a part of the translation business for as long as people who don't share a language have been attempting to correspond. But it was a relatively marginal use of translation until free machine translation on the Internet. The fact that such translation is immediate and free, allows people to try it out without expense. The interactive aspect means that translation errors can be repaired via the dialog.

ECONOMIC AND SOCIAL FACTORS

With the exception of machine translation used for communication and other applications, translation is typically expensive and relatively slow. Translation requires significant planning and coordination on the part of both the client and translator or agency. For these reasons translation of texts has not been undertaken lightly, and the choice of languages has been highly constrained by economic and social factors.

The economic factor is that translation is often justified by either a strong market for the translated material itself, or for a product that the translation will accompany as documentation. Social factors that motivate translation include the needs of multinational organizations (such as the UN, NATO, the World Health Organization), governments of states with more than one official language (Quebec, Canada, and Hong Kong), and finally, the need of governments and businesses to stay aware of developments published abroad that constitute strategic business, political, or military intelligence. So the bulk of translation takes place between the languages used for trade and administration, and those spoken in countries that need to be monitored for strategic reasons.

The languages available in machine translation systems also give an indication of the conservatism in translation. To construct a new language pair in the current state of the practice is a multi-year, multi-million-dollar undertaking. That investment has to be justified either by a significant financial incentive that will underwrite the development cost of machine translation many times over, or by some other social or political motive. In the United States, historically, commercial machine translation developers built systems to translate from English into the languages of American export markets. Industrial clients such as Xerox and Caterpillar have had large enough exports to justify underwriting the development of machine translation systems for their own use. In the other direction, the U.S. government has tended to fund development of translation from foreign languages into English where the foreign language had some significant intelligence value – military, political, economic or related to law enforcement. Two of the earliest such systems were a Vietnamese translation system, developed by Logos Corporation (www.logos-usa.com) for the U.S. during the Vietnam War, and around the same time (which was also during the cold war with the Soviet Union) development of a Russian-to-English system was funded at SYSTRAN Software (www.systransoft.com). More benign examples are the use of machine translation for agricultural and medical texts for use in the member countries of the Pan American Health Organization (http://www.paho.org/english/AGS/MT/), for weather reports in Quebec, Canada (using the Meteo system), and for official communication within the European Commision (using

SYSTRAN.) The result is that well-developed machine translation systems only exist for twenty or thirty language pairs – the languages of countries that constitute rich commercial markets, organizations where a multilingual environment necessitates making official documents and communications available in all member languages, and the languages of countries where conflict, competition or distrust motivate the development of systems for surveillance purposes.

The pattern illustrated by the investment in development of machine translation systems is a mirror of the translation market itself. Literature is not translated unless there is a market that will easily cover the cost of translation, and justify the burden of production and distribution of different versions of a book. Products are not exported, and the accompanying promotional and maintenance literature is not translated unless there is a market that justifies the expense. Scientific articles are not translated out of idle curiosity, but because they contain strategic information that is more valuable than the cost to translate them. The 'surveillance' use of translation is obviously not assessed in economic terms, but in terms of the perceived strategic needs and resources of the translating organization.

FORCES THAT WILL SHAPE THE FUTURE OF TRANSLATION

Globalisation, Internationalisation, and Localisation

Globalisation is getting a tremendous amount of attention as companies realise the potential to do business internationally. The Web gives instant access to any market with a computerised population, and effectively levels the playing field for newcomers and small businesses that would like to sell to a global audience. Additionally, with products that can be sold as 'bits' (publications and software), the problems of shipping and distribution are eliminated because customers can simply download what they buy directly from the Internet. Customers of information products are likely to be comfortable with online self-serve customer support, which may eliminate the need for the vendor to have a physical in-country presence. What is still necessary is to have *localised* pre-sales information, a transaction site that reflects the local currency and legal conditions for online transactions, and customer support. The desire to take advantage of global business opportunities, online and otherwise, is the driving force behind much current interest and innovation in translation.

Let's look at the terms *globalisation, internationalisation,* and *localisation* and consider what they are coming to mean for the field of translation. *Globalisation* encompasses the business strategy and accompanying strategic processes involved in doing business internationally. Ideally, the decision to globalise a business is made very early, when products are still in the design stage.

Internationalisation is a group of procedures and standards that should be followed when designing products that will be sold internationally. For example, any text presentation or editing components should be prepared to work with any of the major character encodings, such as the regionally-used double-byte character sets for Asian languages, and Unicode, which will gradually become the international standard. Additionally, text presentation components should be prepared for non-roman keyboard overlays and the alternative input and editing systems that have been developed for non-roman writing systems, and those read from right-to-left, or top-to-bottom. The core of an internationalised product should make no default assumptions about the representation of currency, laws or contracts that buyers may enter into as users of the product. And finally, an internationalised product should contain no 'hard-coded' English or other strings, for example in error messages.

Localisation is the process of making a product appropriate for the language and culture of some locale. Although localisation is largely a translation process, the experiences of the multinational companies that have led the way in moving from generic translation to the establishment of localisation as a very specialised commercial service that global businesses depend upon to establish and maintain their corporate image, and regional competitiveness. According to LISA[1], the ideal localised product should appear to be produced in the region in which it is sold. This applies to everything from the dialect used in text, to the appearance of the packaging.

---

[1] The Localisation Industry Standards Association, an international professional association of translation and localisation companies and multinational businesses: www.lisa.org

ESTABLISHMENT OF ENABLING TOOLS AND METHODS

The current interest in globalisation and the accompanying product design and localisation issues have spawned an innovative and highly competitive group of software and service providers specialising in the particular needs of multilingual product localisation. This includes significant changes in the authoring and publication process, so that documentation to be translated is typically held in internet-accessible 'content repositories' that can be used as pickup and drop-off points for authors, editors, translators and web-designers. The repositories are automatically monitored for any revisions to the source. When changes are detected, highly automated project management tools extract the revised text, and can send it to translators with the appropriate skills and specialisation, together with any special terminology lists that are to be used with the project. The configuration of such systems is in a state of flux (in 2001), and the industry is characterised by frequent mergers and acquisitions as providers of localisation tools and services try to expand their offerings. This intense development activity has resulted in rather low profitability for the localisation industry as a whole. But the work that is being done now will form the infrastructure for highly efficient delivery of translation for any purpose, not just localisation.

The United States has been notorious for ignorance of translation and related issues, and for the lack of translator training and accreditation programs. Translation has often been viewed as a clerical function, easily performed by anyone who has taken a language class and owns a bilingual dictionary. When there is so little awareness of translation, it has been difficult for translators to get recognition for translation as a skilled trade, and difficult or pointless for even highly skilled translators to get paid on a par with other skilled professionals. While the situation is much better in Europe, where bilingualism and multilingualism are common, and where even the average citizen is a fairly discriminating consumer of translation, there has still been a tendency for translation clients to purchase translation services based on price, with a predictable result in terms of quality.

Now, because localisation clients view the result of translation as the vehicle of their corporate image abroad, they have become aware of quality issues in translation, and are willing to pay a premium to get well-qualified translators and invest in a rigorous review process, in much the same way they would invest in original language marketing or PR materials. The shift in emphasis from price-per-word to quality in the localisation sector of the translation market is now spreading to the rest of the translation market.

With the automated job management processes described above, and with most of the transactions between translators and translation of localisation companies taking place online, employers have begun to recruit translators internationally. However, because of the episodic nature of translation work, translation and localisation services typically employ very few full-time translators, preferring to hire contract translators for piece work. While this approach to business saves the expense of keeping idle employees around on the chance of future need, it also exposes translation providers to two types of problems. First, the need to have a pool of translators to draw on for rapid response to translation requests that are almost always time sensitive. Second, large translation jobs must typically be divided among multiple translators, which is likely to introduce inconsistencies in style, in exactly the market where clients value consistency.

These potential problems are addressed by two trends. First, the compilation of international directories and networks of translators enable employers to identify contract translators anywhere in the world with exactly the skills they need. Second, translator's software tools are in widespread use, and sometimes even required by translation agencies, which facilitate and enforce consistency. The first of these tools is 'translation memory'. A translation memory is a database of pairs (or groups) of previously translated sentences in the source and target language(s). When translation involves text with many repetitions, or revisions of a previous translation, the potential to reuse the translation of the same, or similar sentences, can accelerate translator productivity considerably. These same tools also do the job of identifying the sections that have changed in a revised document. While translation memory has come to be highly valued as a labour-saving device, and a means of providing ever-improving translation quality by re-incorporating final editorial changes in translated texts, it is also coming to be seen as a 'multilingual content database, not just bilingual memory' (Esselink 2000).

The second tool is a terminology manager, which is essentially a bilingual (or multilingual) dictionary with some additional functionality. The terminology manager typically comes without any entries. The user builds up a terminology list based on the terminology prescribed by a translation client, as well as on his or her own topic-specific knowledge. Often translation companies get such terminology lists from the client, which they then copy and provide to each translator on the job, together with any existing translation memory databases, if available. The terminology management programs often work closely together with the translation memory tools, so that for any sentence that has not been translated before (and is not available in the translation memory database), any terms found in the terminology manager will be looked up and plugged into the target sentence before the translator begins to translate.

A final development in this area is widespread adoption of 'controlled language as a means to improve the clarity of documentation, and also to improve the translatability of a text. Controlled language may be as simple as a company-defined terminology list. At the other extreme, it may be a complete definition of the acceptable phrases and sentence patterns usable by the authors of documentation, with special grammar checking software to enforce compliance. The bottom line is that as translation has become so much more important to businesses, even the original language documentation is written with translation in mind.

Translators have often felt threatened by the technology foisted on them by efficiency experts, and by the prospect of losing their jobs to fully automated computer translation. But in fact, the result of the partial automation of the translation process has been to eliminate much of the drudgery and repetition in technical translation. The intense focus on efficiency, quality and best practices has also led to a much better awareness of the levels of skill and investment necessary on the part of translators for different types of text, and an understanding of what types of text may not be suitable for automation at all. For example, promotional and persuasive materials, literature, etc. – the very texts that are most interesting and challenging to translate – are the ones that will not become candidates for automatic translation at any time in the foreseeable future.

In summary, developments in the translation market as a result of the current localization frenzy include:
- Development of many labour-saving tools for translators

- Development and ongoing refinement of a set of tools and methods – the 'best practices' of translation project management
- Compilation of worldwide networks of translators and agencies, resulting in a global labour pool for translators.
- Increased prestige for highly skilled translators
- Growing repositories of already-translated materials (we'll see why this is important in the section on language technology.)

CONTENT DELIVERY AND TRANSLATION ON DEMAND

When most of us think of translation projects, we imagine that there is a text or body of texts at the beginning, and that the set of texts are all translated when the project is concluded. However, with a multilingual website that is a content repository, for example a self-serve technical support site, some of the available texts may be translated frequently into many languages, and other may never be requested for some of the possible user languages. Under these circumstances, it makes sense to have well-written texts that are translation-ready, and provide automatic translation on demand. This is exactly what the Californian CAD software company, Autodesk, has done (Schenker 2001). Support documents that are frequently requested will have polished translation available on file. But most of the 5,000 customer-support documents will be translated on-the-fly only when they are requested. Translation provides a nice example, but provision of many kinds of information on-demand is a way of letting consumers customise and personalise the service and information they receive online, as well as allowing actual use to dictate where the provider invests the most effort.

LANGUAGE TECHNOLOGY

Natural Language Engineering

Machine translation has been under development for over 50 years. Basic research on language has begun to produce a number of other language engineering tools such as spelling checkers, grammar checkers, automatic summarisation, proper name identification, information extraction (identifying the who, where, when of a target event), spoken dialog systems, cross-lingual information retrieval, question answering, and topic detection and tracking (including text visualisation tools). These tools, and the underlying theories and algorithms (procedures for deriving the desired results) are developed in the academic field called 'computational linguistics', which exists at the intersection of computer science, information science (historically related to library science), and linguistics. Within computer science, considerable work has been done within the sub-discipline of artificial intelligence. Until around 1990, much of the work in computational linguistics was 'rule based'. This means that researchers or developers had to devise and codify each rule that the system used to perform its task. A variety of methods were developed to increase the power of these rules, for example using abstract semantic categories to enable some meaning-based generalisations, and formal logic to 'reason'. Without these capabilities, if our rule-base includes the rule 'dogs bark', and we need to determine whether poodles bark, we will have no way to do so unless we add the rule 'poodles bark.' With some semantic abstraction, we can include the information that 'a poodle is a dog', and then be able to infer that poodles bark because dogs bark and poodles are dogs. But the technology was fundamentally limited by the ability of the human programmers and developers to observe and encode all the rules necessary. This is sometimes referred to as the problem of 'scaling up'. Rules can be defined for some well-defined subset of the language, but if the system cannot be efficiently scaled up to handle unrestricted language input. Systems for limited subsets of language are still being developed today, and they enjoy considerable success, for example, the machine translation system called KANT, developed by Carnegie Mellon University for the American earth-moving equipment manufacturer Caterpillar. The documentation for the equipment manufactured by Caterpillar can be automatically translated, and requires little or no post-editing, because the subset of the language handle by KANT is so limited. The system has lots of world knowledge that has been carefully encoded into it. But it doesn't need to know about anything but tractors.

AUTOMATED LEARNING

From the mid 1980s, and more strikingly from around 1990, statistical methods of language analysis and modelling, and ultimately translation, were developed (Brown, Cocke et al. 1990).[2] These methods enabled researchers to write programs that could learn how to translate between languages merely by analysing large amounts of aligned translated text. For example, in addition to automatically learning which words or phrases of French were typically translated into which words or phrases of English, techniques were developed and refined which allowed the automatic learning of 'language models'. These models of single languages allowed a computer program to assess how well-formed a candidate sentence is compared to other candidate sentences (or rather how 'likely' it is.) The point here is not to delve too deeply into statistical natural language processing, but to illustrate the point that a variety of techniques emerged, and have been further developed and refined over the last 10 years, which have the potential to do the same work as the rule-based approaches, without the investment of human labour.

Earlier, I mentioned that there are limitations on the types of texts that are suitable for automatic processing, and that human translators will always need to be involved in producing the highest quality translations. One of the reasons that the taxonomy of uses of machine translation (*assimilation, dissemination,* and *communication*) are well-developed and understood, is that the 'success stories' of machine translation implementation have been limited to those situations where:

- Speed is more important than quality (assimilation). Translation is valuable even if the user can only get a gist of the translated text, and determine whether or not to have a human translate it.
- Quality is important, but the user is writing the text and can do so in a way that maximizes the chances of high-quality translation (dissemination).
- The translation happens in real time as part of a dialog, and errors can be overcome by rephrasing or other conversational devices (communication).

Even within these applications, only the best-written, clearest texts come out very well. Partly, this is due to problems handling ambiguity in text, and partly it is due to limitations in the coverage of linguistic patterns, words, and multi-word terms. Statistical meth-

---

[2] Two recent textbooks provide overviews of the developments in this area in the 1990s, (Jurafsky and Martin 2000) and (Manning and Schuetze 1999).

ods have the potential to do somewhat better in handling ambiguity by their ability to make purely quantitive observations on how words are used and translated. They also have the potential to greatly extend the coverage of translation systems. Whereas a human linguist may use 100 or even several hundred examples in developing translation rules, a statistical system will learn from millions of examples.

Even after 10 years, statistical methods are limited by the fact that they require tremendous amounts of previously translated texts in order to learn how to translate a language pair. However, they will be refined over time, and will eventually be able to learn how to map one language to another with a smaller amount of data. The importance of statistical methods is in their ability to automatically learn rules from language data, and in their ability to produce more natural sounding translations than those produced by any other method. Statistical methods, in the right combination with other approaches hold out the promise of producing usable quality translations from many more types of text than have previously been translatable by machine.

NATURAL LANGUAGE UNDERSTANDING

One of the most intractable problems for natural language processing is the ambiguity of language. A sentence such as, 'I saw the man with the telescope' is hopelessly ambiguous in a way that cannot be resolved without actually witnessing the situation described. Did the man have a telescope? Or was it through the telescope that the speaker saw the man? Sentences like this, which include ambiguities that come to our attention, are relatively rare. However, for computers, which lack 'world knowledge' most sentences are highly ambiguous. If we step back from our knowledge of how the world works, we can see that the sentence 'Time flies like an arrow.' can have at least two interpretations. In the obvious interpretation, it is a metaphorical way of describing how quickly time passes. In the other interpretation, there is a kind of fly, called a 'time fly'. It so happens that these flies like arrows. A human reader may be baffled and appalled that such a ridiculous interpretation is possible, but it is characteristic of the types of errors made by computer analysis of language. Avoiding and recovering from this type of ambiguity occupies much of the language engineer's effort in traditional rule-based system development. Are there better ways of tackling ambiguity and the world knowledge problem?

The statistical approaches mentioned above are sometimes called 'shallow' because they look only at the surface form of language. They often do not attempt to analyse any grammatical structure in texts. In parallel with the 'shallow' statistical approaches, other research paradigms have aimed at making automatic analysis of the meaning of language possible. These efforts, sometimes called 'deep' or 'knowledge based' approaches, have invested heavily in the development of taxonomies of the world, or semantic knowledge bases. Some examples are Wordnet,[3] CYC,[4] and the Universal Networking Language project.[5] The ultimate goal of these taxonomies is to give computers world knowledge that will allow them to understand natural language, and reason about the events described in language. More modest and practical goals for the same resources are to make it possible to annotate text for meaning in a hidden layer of representation, in much the same way that text can be annotated for format using HTML tags. Historically, knowledge based approaches have been coupled with rule-based methods, but researchers in the statistical learning paradigm are now looking to the future and are considering utilising text that has been annotated for meaning to 'learn' the meaning of text automatically.

Some machine translation projects have utilised a combination of semantic knowledge bases and deep grammatical analysis to come up with 'interlinguas': language-independent encodings that allow a complete representation of the meaning of a text, without the ambiguity of natural language. Interlingual machine translation systems perform a deep analysis of the source text in order to extract the full meaning of the text. While many projects have attempted this, it is actually very difficult to encode the meaning of language without actually using an existing natural language, and without biasing the encoding towards the grammatical structure or worldview of the languages or cultures that such projects begin their work on. While people have developed complete and unambiguous notation systems for representing math and music, natural language does not

---

[3] www.cogsci.princeton.edu/~wn/
[4] www.cyc.com
[5] www.undl.org

lend itself easily to this kind of codification, and true, general purpose interlingual machine translation has proven hard, if not impossible. However, as we will see below, there are big rewards if we can succeed at it.

TRANSLATION AND MULTILINGUALISM IN THE FUTURE

Throughout this chapter, we have repeatedly referred to machine translation. While the technology has been available in a rudimentary form for several decades, it has not been of high enough quality, or flexibility to handle most types of text well. At the same time, while the volume of text translated by human translators has increased continuously and much of what might usefully be translated cannot be translated, either because there are not enough translators to do it, or because the text has such a short 'shelf life' that by the time a human could process it, it would have lost its timeliness. Many of the texts that people would like to have translated will not be suitable for machine translation, but if machine translation were capable of taking on more of the easy and time-sensitive jobs, human translators could focus on the harder jobs.

It is a sad fact that many of the world's languages will be lost. Probably half of the world's current 6,000 or so languages will shortly die out from lack of use.[6] For the languages with only a few surviving speakers, probably nothing can be done. For many others, it might seem that an increase in translation would be a good start in addressing the problem. Wouldn't this allow developing countries to join the online world, and interact in their own languages? Might it at least allow such people to participate in the global economy and benefit from the educational resources available in other languages? Might it reduce the pressure to abandon native language for English, or some other lingua franca of trade?

To some extent, this may, and probably will happen, though it will probably work for far fewer languages than we would hope. Here are three obstacles:

- Unwritten languages: for the many languages that are not written (and hence have no accompanying written tradition or literacy

---

[6] SIL, Ethnologue estimate. www.sil.org

education), there are many practical and cultural steps that would have to precede making translation available.

- The N^2 problem: The Anglophone world may presume that all translation is either into or out of English, but if we really want to promote international communication, commerce, and resource sharing, we should think in terms of translation between any two languages. But what would this mean? If there were just 5 languages in the world, a complete exchange of one text from each of the languages (translating one text from each of the languages into all of the other languages) would involve 20 translations. If we were developing machine translation systems to do this work, we would have to build 20 language pairs. The formula for this is $N \times (N - 1)$, where N is the number of languages. Computer scientists tend to simplify this, and describe the formula as N^2. Even if we decide that only 1,000 languages are suitable for regular exchange of information via translation, we are faced with 1000^2 or 1,000,000 language pairs (or, strictly speaking, 999,000 pairs).

-

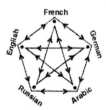

Figure1: Dirst translation between all languages. What a mess!

- Translation, or development of automatic translation systems is currently expensive, and small speaker populations are unlikely to attract the investment. However, committed communities of speakers, together with some source of funding can do the work. One current example is the CATANAL project, to save and perpetuate Alaskan native languages via the development of computer-aided language learning and translation software. (A report on the project is available at http://hometown.aol.com/ in the path mit2usa/Partners.htm under Peter Wilkniss.)

The problem of saving endangered languages, and bringing languages at risk into the mainstream is one for public policy, and for speaker communities. Market forces will not address these issues. However, the developments described above in both the translation

119

business, and natural language technology may make it possible to make most texts from any language available in any other language, and thus slow or stop the loss of the world's linguistic diversity, as well as opening new markets for published information which would formerly have been considered too small to bother with.

The N^2 problem is greatly alleviated by the interlingual approach to machine translation, described above, or the 'pivot' approach. If text to be translated could be converted into some alternate representation, from which it could then be converted to the target language, the number of translation components would be only $2N$, or 2,000 for our hypothetical 1,000 languages for translation. This is a much more manageable set of components, which could allow translation between all 1,000 languages. In the section on natural language understanding we described interlingual translation systems that use an abstract representation as a way to store the meaning of a text. However, some machine translation systems, typically called 'pivot' systems, use a natural language as the interlingua, for example English or Esperanto. This approach could still work with surface-only statistical methods, avoiding the human effort and scaling up problems of interlingual approaches.

Figure 2: Translation using a pivot language

As mentioned earlier, statistical approaches to translation typically learn direct translation rules between languages, and interlingual approaches involve a deeper analysis of the source language through a set of manually developed rules (followed by additional rule sets to select correct meanings in the target language and then generate the target language text.) It might be possible, however, to hybridise the approach, using statistical methods to learn a mapping between the source language and the pivot language or interlingua, and then again between the pivot or interlingua and the target language. It is possible, even likely, that the shallow statistical pivot method of doing machine translation will not be powerful enough to enable moderate quality general purpose translation. This suggests an analytically deeper approach. However, I am pessimistic about the chances for interlingual methods to fill the gap because of the human labour involved, and the difficulties in scaling up such systems. I talked earlier about the development of semantic category systems that could be used to annotate texts for some aspects of meaning. It may be that some sort of light markup for meaning will provide enough additional power to allow statistical methods to yield moderate quality translations.

Development of translation components using the pivot approach, where bilingual texts are available for training, (for example utilising existing translations of the Bible which are available for virtually every written language), could be relatively fast and efficient once the development method has been sufficiently refined. The cost of developing the method will be eagerly subsidised by businesses in the big languages of international commerce. Economists refer to the distinction between fixed costs (the cost to produce the first copy of a product or service) and marginal costs (the cost of each additional copy/iteration of the product or service). For information goods, the fixed costs are typically high, involving many years of effort and resources, but the marginal costs of reproducing the information is negligible, the cost of a CD or a download (Shapiro and Varian 1999). The development of an automated system for learning translation rules should be similar. Years of time and effort will be required to develop and refine the methods and computer programs. But once the methodology is complete, repeating it for new examples – new languages – is essentially free.

The confluence of research, experimentation, and solid progress in many areas related to translation will make automatic translation inexpensive and accessible to any written language in the future. How soon this happens depends on the ongoing investments in R&D – in the commercial world, where research money is in turn dependant on the continuation of a reasonably strong global economy and increasingly free international trade; and in the scientific research community where funding is dependant on the social climate and priorities of funding agencies. The ideal of usable quality automatic translation capability for all written languages is some distance away from us – at least 5 years, if not 25 years. But for the world's major languages, the capability is essentially already available. Exploiting best translation practices, adopting writing styles that facilitate translation, and putting pressure on the developers of machine translation and translation tools to improve their offerings, will put near-publication quality automatic translation within reach in the very near term.

Chapter 9

SKILLS DEVELOPMENT FOR MULTILINGUAL DIGITAL PUBLISHING

Michael Singh

The development of the publishing and printing industry as a medium of mass communication contributed to efforts by nation-states to consolidate the notion of languages defined by national boundaries, and nations defined by languages. Thus began the magnification of the inequalities between the world's languages and the erosion of humanity's multilingual knowledge base. Current innovations in second-generation digital technology occupy a contrary space; they are now facilitating the efforts of authors, editors, publishers, printers and booksellers to open up the possibility of a truly multilingual knowledge economy globally. With cyberspace blurring the boundaries of nation-states, issues of multiculturalism and multilingualism matter as much as ever. Multilingualism is a fast expanding field for those electronic and digital publishers and printers seriously interested in carving out a niche in the international marketplace. Bilingualism, or proficiency in a second language, is now a valued added resource for ethical and active shareholders keen to invest in the multilingual knowledge economy at the same time as contributing to the sustainability of the world's linguistic diversity (Rose, 2001). The emerging role of ethical investment means that shareholders are making choices. They can now choose between those companies which use second-generation digital technologies to help sustain linguistic diversity rather than those who, as David Crystal (2000) reports are contributing to the global dominance and killing power of English (and several other global languages such as Mandarin, Spanish and Arabic).

After locating multilingual book production using second-generation digital technology within the context of the multicultural and multilingual attributes of Australian industry and its workforce, this chapter canvasses several features that are worth addressing when framing a skills development program that supports the education of a multilingual workforce. This chapter considers four features useful for framing skills development programs that are supportive of the linguistic diversity now needed for re-positioning Australian industry in the multilingual knowledge economy. Four key dimensions of a skills development curriculum framed to support productive linguistic diversity are: adding value to cultural and linguistic differences; providing intellectually challenging learning experiences; building real world knowledge connectivities; and providing a supportive teaching/learning environment (Lingard, Mills and Hayes, 2000). This chapter is structured to provide an outline of issues relating to the development of multilingual book production technologies:

- What are the multicultural and multilingual features of Australian industry and its workforce?
- How can education and training add value to Australia's cultural and linguistic diversity?
- What intellectually challenging learning experiences support the preparation of workers for the multilingual knowledge economy?
- What real world knowledge connectivities can be built between students' learning experiences and their future career trajectories?
- What provisions are needed to create teaching/learning environments supportive of the inter-generational transmission of languages?
- What processes are contributing to the public education of investors about issues of corporate responsibility?

WHAT ARE THE MULTICULTURAL AND MULTILINGUAL FEATURES OF AUSTRALIAN INDUSTRY AND ITS WORKFORCE?

For Australian publishers and printers to participate in the global multilingual knowledge economy they need to develop, or have access to, the capacity for issuing books in local languages suited to local conditions, or otherwise they may deny themselves significant market opportunities. The viability of the local publishing and printing industry in Australia lies in its ability to actively engage in the global multilingual knowledge economy. The industry has an important role to play in knowledge transfer from one country via language to another, and thereby helping to build up the knowledge resources of other countries (Smith, 1977). The aims of the Australian industry include increasing the size of its potential markets; improving its return on investment; providing shareholders with ethical investment opportunities and to embellish the image of individual companies and the Australian industry as a whole.

One possible means for doing this involves having due regard for language sustainability through productive marketing advantages of cultural and linguistic differences. Moreover, the production of books containing knowledge and languages from localities around the world undermines the charges of reinforcing knowledge dependency and linguistic imperialism that frequently arise from the one-way flow of knowledge. Many local authors from around the world would find considerable prestige in having their work published in their own language by a reputable and socially responsible transnational company operating out of Australia.

The geographically and linguistically protected regions of the book trade constructed around English are disintegrating under the pressure of an open market. Therefore, it may be useful for the Australian publishing and printing industry to investigate how to make the nation's diverse cultural and linguistic knowledge base a productive advantage in the multilingual knowledge economy. To advance the economic potential of Australia's multicultural and multilingual resources this may even mean, in some instances, working to mitigate the long-standing mindset of publishing and printing English-only products for protected English-only markets. This mindset is likely to hinder the development of an industry that could otherwise make a competitive advantage of the global business and social networks of Australia's multicultural and multilingual workforce and business community.

Jason Epstein (2001) was awarded the Curtis Benjamin Award of the American Association of Publishers for inventing new kinds of editing and publishing. Epstein has challenged this mindset. He argues that traditional publishers have an emotional and financial attachment to the established economic models that create difficulties for them in making the transition to the production of 'a vast multilingual virtual library' with audiovisual and interactive enhancements. Here is an opportunity for the primary producers – bilingual authors and (small) businesses, publishers and printers to meet the need for independent Australian books and to tackle the world's multilingual book markets. Given that the USA, UK and Canada have yet to remove territorial copyright, the means they use to protect their restricted markets, this makes it difficult if not impossible for Australian publishers and printers to sell into those large English-reading markets. This suggests the desirability of investigating a range of multilingual markets as a way of securing a competitive advantage in an open market where the Australian industry will now have to compete with US and UK books in English.

While a significant market for ebooks is yet to emerge, the online distribution of digitally produced books is likely to become a key element in future book production. Dictionaries, atlases, encyclopaedias, directories and glossaries are generally suitable for e-book production. Jason Epstein (2001) reports that, given the availability of electricity, paper and the requisite printing press, there is the capacity for books on digital files (dbooks) to be printed and bound at the point of sale by machines – in supermarkets, photocopy shops, coffee shops, schools, libraries or student residences.

Book retail pricing was deregulated in Australia in 1972. The 1991 amendments to the Copyright Act resulted in some increased competition from parallel imports and some decreases in prices. The hike in cover prices created by the GST and the decline in the buying power of the Australian dollar in 2001 is now compounded by the industry having to face the unilateral opening of the Australian book selling market to international competition (Steger, 2001). The latter is being achieved by the removal of parallel import restrictions via the Copyright Amendment (Parallel Importation) Bill 2001.[1] These developments provide challenging opportunities for the local publishing and printing industries, and everyone associated with the local production of intellectual property.

For publishers, distributors and booksellers importing large amounts of books from overseas, the possibility is that this could mean cheaper books for consumers. The creation of an open market could also mean an increased disincentive to invest in local authors, the dumping of remaindered books from overseas, as well as undermining the capacity of Australian authors to earn royalties. It could also mean the loss of 400 jobs, a cut in business of $40 million and the failure of some individual publishing and printing companies (Steger, 2001). However, given that small print runs for

---

[1] see www.asauthors.org – click on the 'parallel imports' link to see what publishers Michael Heyword and Patrick Gallagher have to say

Australian editions could become too small for overseas publishers to justify, this may open up innovative, globally-oriented publishing and printing opportunities for the Australian industry.

One of the economic benefits of Australian multiculturalism lies in the diasporic connectedness of citizens who have business and linguistic links to various places and markets around the globe, including those that are intensifying and accelerating Australia's linkages with Asia, the Middles East and Latin America. In the global multilingual knowledge economy, Kotkin argues for:

> ...the acquisition of knowledge and cosmopolitan perspectives ... In an ever more transnational and highly competitive world economy, highly dependent on the flow and acquisition of knowledge [in different languages], societies that nurture the presence of [their multicultural and multilingual resources] seem most likely to flourish (Kotkin, 1992).

Once, it was the English diaspora that linked the English-only publishing and printing industries of Australia, Canada, the UK and the USA. But the ethnically and linguistically protected links to these particular regions of book trade are now disappearing. Now it is the diasporic connections of Australia's diverse cultural and linguistic knowledge base that is important for international trade and investment across a range of Australian industries. It is the ability to take productive advantage of this knowledge base in order to nurture export markets and increase trading opportunities that has encouraged businesses such as Writescope to re-imagine the links between Australia's multilingual resources and their printing and publishing efforts. The opening up of the international book trade creates possibilities for negotiating multilingual, transnational linkages. For an innovative digital printing and publishing industry now dependent on new niches in the international market multilingual products and services, second-generation digital technology is now available to enable them to target the linguistic particularities of a diverse global marketplace. The industry can draw on the language resources of multicultural Australia to provide necessary authors, editors, proofreaders, translators, web managers and cross-cultural sales representatives.

In terms of the business supply chain it is necessary to consider the potential workforce available for the production of books for globally dispersed multilingual markets. At the time of the last census in 1996 there were approximately 2.5 million Australians who between them spoke some 200 different languages. This gives

Australia a competitive advantage in building links into the multi-lingual knowledge economy (Australian Bureau of Statistics). The five largest languages in addition to English included Italian (367,300), Greek (259,000), Cantonese (190,100) Arabic/Lebanese (162,000) and Vietnamese (134,000). These figures should not be read merely in terms of possible Australian niche markets, but rather seen as representing the potential of the Australian workforce to be involved in producing goods and services for a multilingual knowledge economy.

One of the important benefits of Australia's post-Second World War migrant heritage is that Victoria is helping to extend this multilingual skills base by offering forty-five languages for students to study at the VCE level (Ashenden and Milligan, 2001, p. 8). In the year 2000, some 9,000 students studied VCE languages ranging from Chinese, French and Japanese, Albanian, Lithuanian, Slovenian, Latvian and Estonian. There are more than 200 community language schools throughout the State teaching some 50 languages to approximately 30,000 students who attend after-school or weekend classes. Bilingual education programs are also provided in at least 10 languages including Arabic, Greek, Indonesian and Vietnamese. Among the range of careers where these languages can be useful and which relate specifically to the publishing and printing industry are those of editor, interpreter, proofreader, translator, exporter and importer, market researcher, sales representative, publicist, website manager and writer.

Whether sufficient government investment is being made in producing a multilingual workforce to meet the needs of industries keen to participate in the global multilingual knowledge economy is a question that requires further investigation.

HOW CAN EDUCATION AND TRAINING ADD VALUE TO AUSTRALIA'S CULTURAL AND LINGUISTIC DIVERSITY?

Consider for a moment the following story, which is indicative of the added value to be gained from linguistic diversity. A troupe of 29 Navajo radio operators who used their language as a means for transmitting secret, coded messages during World War 2 is credited with helping the USA military win the crucial battle for Iwo Jima (Campbell, 2001, p. 15). While the complex syntax coupled with the guttural, nasal pronunciation of Navajo makes it extremely difficult for non-speakers of the language to decipher, its use in the war reduced the transmission of messages from 30 minutes to 20 seconds. Moreover, faced with the challenge of *bringing their language forward* to meet the requirements of changed circumstances, these Navajos proved that their language had the capacity to convey contemporary ideas about the war: bombs, grenades, battleships, submarines and such like. US President Bush awarded the five surviving members of the unit the congressional gold medal in recognition of their service. Their story is told in a John Woo film entitled *Windtalkers*. Here it must be noted that prior to the war these Navajos were punished for speaking their vernacular in school. No doubt the loss the Navajo language and its knowledge could have meant many more US deaths than the 7000 American troops who died in February 1945.

A skills development program framed to support productive linguistic diversity adds value to Australia's cultural and linguistic knowledge base, and pro-actively seeks to increase the participation of students from different cultural and linguistic backgrounds in the multilingual knowledge economy. In doing so, such a skill-building curriculum explicitly and deliberately add values to the transnational cultural and linguistic communities, of which students are a part. Moreover, students can learn to draw funds of knowledge from these communities to contribute to the re-positioning of Australian industry within the changing global marketplace (Lingard, Mills and Hayes, 2000).

This 'adding value to cultural and linguistic diversity' calls for the acknowledgment of inequities in the sustainability of different languages, rather than flattening out and making structured differences between humanity's 6,000 languages merely seem equal for the sake of appearance. This means recognising that some languages have a greater chance of long-term sustainability in relation to others. This is because some languages have more power and prestige than others. It is also because the range of resources and choices available varies according to factors of history, population size or the undervaluing of other languages, their speakers and their knowledges.

While we all speak our particular language from a particular place, history, experience and culture, no language is forever contained or constrained by that positioning. In this context there are real possibilities for the printing and publishing industry to have a role in enhancing the sustainability of world's linguistic diversity and its associated knowledge base.

In cyberspace, linguistic differences have not been rendered invisible or irrelevant. On the contrary, they are now more important than they were during the first generation of digital technology. Rather than eliminating 60–90% of the world's different languages in the course of this the 21st century (Crystal, 2000; Nettle and Romaine, 2000), second-generation digital technology has now made it possible for many of these language to count in the emerging multilingual knowledge economy. In a program of skill development framed to support productive linguistic diversity, it is made to count by enhancing students' learning of a second language in addition to English. This adding of value to cultural and linguistic differences, through recognising and incorporating it into the curriculum, is allied to the development of students' higher order thinking skills such as critical analysis.

The dominance of English in the first generation of digital technology was not just a function of its colonialist history, but also because of its claim to offer valuable cultural or intellectual capital. The second generation of digital technology has emerged because of the need to access the world's linguistically diverse marketplace and growing recognition of the ways in which English-only products limit the development of the world's multilingual knowledge economy. There is also growing recognition of the importance of the intellectual capital available in other languages and the desire by speakers of those languages to record and circulate that knowledge in their own language. We all know what a loss there would be if all the English language books in our libraries were burnt; but what loss of collective human knowledge are we allowing to be destroyed through the extinction of the world's languages and the associated death of the knowledge they contain.

Language is what makes it possible for human beings to survive across the generations in different places on earth; making it possible for us to create and recreate culture, technology, art, music and so on. Language is a rich source of accumulated knowledge and human wisdom, a living repository of knowledge and a vehicle for conveying knowledge needed for sustaining the future of every human culture. Where the knowledge available in languages 'othered' through English (and other dominating languages) are not always valued as a human resource, Nettle and Romaine (2000, p. 16) argue that the 'next great steps in scientific development may lie locked up in some obscure language in a distant rain forest.'

One-quarter of prescription medicines in the USA are derived from the world's rainforest plants – knowledge of these gained from Indigenous peoples and their languages. Remember what Les Hidens, the 'Bush Tucker Man', learnt from Indigenous Australians? Just how much more could have been learnt if so many Indigenous peoples and their languages had not been destroyed we will never know. What have we yet to learn about climate or the management of marine resources from the speakers of other languages now that we need to redress our over-fishing and degradation of our waterways?

Not only are languages important for encapsulating human identities and providing historical interpretations of human existence, but they provide a wealth of human knowledge. The sustainability of humanity's language diversity is important to the world's multilingual knowledge economy. What valuable knowledge did the USA lose through its efforts to wash Indigenous languages from the tongues of Native Americans? What precious knowledge of holistic environmental sustainability has Anglophone Australia lost through its efforts to silence Indigenous Australians and their languages? Those few Indigenous languages which have survived have taught Anglophone zoologists and botanists to distinguish between species of pythons, kangaroos, wallabies, insect larvae and plants they otherwise thought were the same, using Indigenous taxonomies based on movement, rather than static appearance (Crystal, 2000, pp. 44–54).

Teaching and learning the diversity of languages already present throughout Australia would increase the human resources available for Australians to engage in the multilingual knowledge economy. These languages now need official status as a national asset, with investment by State and Federal governments in their own continuing linguistic development and the expansion of the numbers of Australians who can speak them. Australia could promote the viability of its diverse language-base by rewarding the speaking, writing, teaching, publishing and 'bringing forward' of these languages, as well as the development of knowledge in these languages.

WHAT INTELLECTUALLY CHALLENGING LEARNING EXPERIENCES SUPPORT THE PREPARATION OF WORKERS FOR THE MULTILINGUAL KNOWLEDGE ECONOMY?

There are four important features of the intellectual quality of a skills development program framed to support productive linguistic diversity. These are; the promotion of higher order critical thinking; acquiring a depth of knowledge and understanding; treating knowledge claims as problematic and open to question, and learning the meta-language framing any given operation (Lingard, Mills and Hayes, 2000). These features have been long associated with student-directed learning, interactive teaching and multidisciplinary real-world project work. Intellectually challenging skills development experiences are usually made possible through what is called 'three dimensional teaching'. This is an orientation to skills development that enables students to become proficient in the related dimensions of the 'technical' arts, basic 'cultural' interpretation and 'critical' skills (Lankshear, Snyder and Green 2000, 42).

This is a well-established holistic view for framing skill development curricula that support the education of a multilingual workforce. Key features of three-dimensional teaching are briefly summarised below. This framework for skills development focuses on enhancing students' abilities across three interdependent dimensions: the mechanical or procedural; the making of good sense in and about social, cultural and economic contexts; and the creation of innovations through critical reflection, analysis and evaluation.

One of the important dimensions of skills development is the transmission of the knowledge and skills students need for doing many different kinds of procedural tasks. This 'technical' dimension of skills development focuses on students learning functional abilities as well as knowing how to use different operations to engage in real-world, work-related tasks. For instance, given the rise of the global multilingual knowledge economy, it is important for students, in the process of their skills development, to gain the 'how-to' knowledge concerning new governmental techniques for intervention and control. This includes developing their 'know-how' with respect to regulation by global market forces, open competition, cost reduction, quality assurance, client responsiveness and enforcement:

> Governments continue to move toward the deregulation of every service area and to embrace competition as the means by which costs of delivery of public services are reduced and quality improvement and respon-

siveness to changing needs are enforced... They seek therefore vocational education, fostering innovation and low tax rates. The Australian government continues as well to focus on opening Australia to world competition (Dunkin, 2000, p.5).

As implied in the cultural and critical dimensions of skills development discussed below, there are certain technical skills that students need to develop in these dimensions as well. They need the 'know-how' to be able to *evaluate* the pace of technological and economic change in relation to people's capacities to engage it; to undertake socio-political and cultural *impact analyses;* and to *monitor* differential levels of anxiety, alienation and disaffection. Students also need to develop the operational capacities required to *program manage* people's engagement with fast moving, radical change; and to *build scenarios* for the development and maintenance of a global civil society. They need the technical skills to enable them to *critically reflect* on government interventions, the opening of the nation to world competition, the morality and politics of an ethic of efficiency, and the political incitation of fear and instability – problems for businesses trying to operate in transnational markets.

The second important dimension of preparing students for the real world of work is the transmission of the knowledge and skills they need to contribute to the development and maintenance of a civil global society and the fulfilment of human life. As part of their skill development, students need to learn to undertake analyses and make interpretations of the *socio-political and cultural impact* of 'globalisation from above' and 'globalisation from below' (Falk, 1999). In some instances the meaning given to the impact of 'globalisation from above' has found expression in revenge politics, which in Australia's case has incited racist fears directed against Indigenous and Asian-Australians. This has made it unnecessarily challenging for Australians conducting international business. In some instances, the social alienation and disaffection this has caused has created instability in Australia's international trading relations:

> We recognise that the pace of technological and economically-driven change that can be accommodated by people and groups is more limited. Alienation is a significant social phenomenon as groups within society are affected disproportionately and face mixed futures. Social problems and disaffection grow in many developed societies. The attractiveness of politicians who articulate the fears and anxieties of people facing such radical change has been clearly shown in elections recently in countries such as, Australia, UK, USA and Germany (Dunkin, 2000).

133

If the Australian printing and publishing industry is to reposition itself within the global multilingual marketplace then it would be useful if its prospective employees appreciated the social and cultural impact of the globalisation of English. As an integral aspect of their skill development it is important for industry workers to know that in potential markets English is regarded as a 'cultural nerve gas' (Crystal, 2000, p. 78), which is seen to be infiltrating everywhere with reinforcement from Anglophone governments and mass media. That the production of English-only products and services creates antagonism among speakers of other languages creates a context for business decisions which favour the development and rejuvenation of languages other than English (Macedo, 2000). With reference to one of the world's worst cases of language extinction and knowledge death, it is important to quote David Crystal (2000, pp. 87, 102, 136) at length:

> In Australia, the presence of English has, directly or indirectly, caused great linguistic devastation, with 90% of [Indigenous] languages moribund... Throughout the 1990s, there were several reports of language support programmes becoming endangered through budgetary cuts... in Australia, at the end of 1998, when the Northern Territory Government announced plans to phase out bilingual education for its Aboriginal communities, replacing this by English-teaching programmes... If their only experience of speech and writing in school is through the medium of the dominant language, it will not be surprising to find that the Indigenous language fails to thrive.

The third important dimension of skill development involves developing in students the knowledge and skills needed for practical innovations. This can be done by nurturing students' critical understanding of the relationships between the cultural fabric of global/local society and the world's changing multilingual knowledge economy. The 'critical' dimension of skill development focuses on engaging students in learning to assess, judge and evaluate the technical, political, economic and socio-cultural aspects of globalisation and localisation in order to create innovative products and transformative practices. This includes developing innovations in methods or technologies, as well as making innovative moves with respect to existing practices and changing the rules governing these practices.

As indicated in the following statement, it is critical reflection on the gaps between hope and happening, and critical analysis of the contradictions between the techniques of 'globalisation from above' and its localised socio-cultural impacts that provides the scope for building innovations:

> The picture of change that we acknowledge, then, is one in which some facets of our society are moving fast and challenging existing models and paradigms. At the same time, we accept that people's capacity to adapt to such change is more limited. Our task is to help them bridge that gap. It is also the challenge for us as a community within RMIT University to ensure that, as an institution that wishes to continue its current role in developing and maintaining civil society, it adapts to an appropriate pace to these external challenges (Dunkin, 2000, 5).

Together, industry application of second-generation digital technologies and governmental investment in the development of the skills of its multilingual workforce represent the major dynamic forces in positioning Australia to make an advantage in the global multilingual knowledge economy. This being so, then it is to be expected that innovation and growth in the Australian publishing and printing industry is to be found in efforts to offer its customers throughout the world all the possibilities for communicating in books in their own languages. For most people throughout the world, the ability to speak at least two languages is taken for granted. It is only the nation-building projects of the last two centuries that have mistakenly equated cultural and linguistic homogenisation with modernisation. This project failed; there are only several hundred countries in the world but there are still (for the moment at least) some 6000 languages. Increasingly, classificatory schema constructed to organise accounts of online use by languages also recognise the failure of this project, in so far as they acknowledge that people speaking the same language can do so from whatever country they happen to live in. The iconographic use of national flags to represent the languages in which a site may be accessed is now inappropriate. This is especially so given that a 'national language' may be spoken in different countries, and moreover, that 'national languages' hide many other languages (and therefore markets) within their borders.

A multilingual publishing and printing industry using se   cond-generation digital technologies could make for more cost-effective and widespread distribution of knowledge than is possible by first-generation, English-only technologies.

WHAT REAL WORLD KNOWLEDGE CONNECTIVITIES CAN BE BUILT BETWEEN STUDENTS' LEARNING EXPERIENCES AND THEIR FUTURE CAREER TRAJECTORIES?

Productive skills development programs enable students to integrate knowledge from across diverse fields and to connect what they are learning with their own experiential knowledge and future life trajectories. Connecting their learning to real-life contexts and engaging in real-world problem solving are integral features of any skills development program framed to support productive linguistic diversity (Lingard, Mills and Hayes, 2000). Efforts to frame such a curriculum involves building these attributes of knowledge connectedness, including maximising opportunities for and the authenticity of 'real world' learning (Lankshear, Snyder and Green, 2000, 118).

Students' skills development can be connected to real-life contexts which show the use of second-generation digital technology to facilitate the publication and printing of books in increasing numbers of the world's written languages. This includes the languages spoken by small numbers of people as well as varieties of English (Gupta, 2001, pp. 148–164). By the late 1990s, it was possible to produce electronic and digital publications in French, many of which came from Quebec, even though French Canadians represent only 5 percent of the world's speakers of French (Oudet, 1997). Speakers of Maori and Navajo are using the second-generation digital technology to build connections among widely dispersed community members and for the low-cost publication of language materials as part of their commitment to language development and revitalisation.

In Hawaii, second-generation digital technology is being used to help preserve and strengthen the local Indigenous language, ka 'olelo Hawaii', after having been forcibly repressed for some eighty years by the imposition of legislative bans by the US/American colonists (Warschauer, 2000, p. 159). Rather than relying on stand-alone texts, multiple media are important in the passing on of knowledge and language among Hawaiians. Not surprisingly the multimedia functionality of second-generation digital technologies is proving to be a more culturally appropriate media for a variety of language communities.

Students learning to become interpreters and translators can be connected to the emerging possibilities related to their future career trajectory through case studies, field projects and internships with businesses such as Charlie Baxter's eTranslate Inc. This business was developed not to serve those collapsed business models which did not incorporate 'going abroad' in their strategic planning, but targeted big companies that had already established international operations (*The Age*, 2001, p. 9). eTranslate provides multinational companies actively engaged in contemporary processes of globalisation with the strategic means of translating company documents; this includes 'localising' them for particular international market segments.

In recognition of the fact that international business constantly requires good translation, major investors in eTranslate include General Electric, the Goldman Sachs group and Bain Capital. The company draws on an internet-based global network of skilled, reliable translators from across multiple countries. While digital technology has produced many innovative products, including software to automate selected tasks involved in managing web content in multiple languages, it has not as yet conquered the fine art of translation. After appropriate training, which includes studying the risks of doing business in a given country, the translators render English language publications into different languages. No company, except maybe a funeral parlour, wants their publications in colours, symbols or numbers to be read as signifiers of death in another language or another culture.

Machine-aided human translation is characterised by providing a brief and concrete abstract of the full text in a machine-friendly manner, so that the abstract can be translated into many languages without losing fidelity. This process involves the analysis of the components of the text, clarification of the text with the author whenever necessary and then translation into an intermediate abstract representation – the gist of the text. This is then used to generate translations in different languages. Computer supported human translation involves:

- having a human translator to correct and validate the machine translated text;
- taking culturally specific ideas into consideration when translating material between culturally different linguistic areas;
- processing files against terminology databases to ensure that technical phrases are translated in domain specific ways;

- using searchable concept-based terminology resources and the-sauri;
- integrating machine translation and domain specific terminology sets with authoring tools to speed up translation services.

WHAT PROVISIONS ARE NEEDED TO CREATE TEACHING/LEARNING ENVIRONMENTS SUPPORTIVE OF THE INTER-GENERATIONAL TRANSMISSION OF LANGUAGES?

Giving students an increasing measure of control over their learning, providing them with positive social support, and ensuring they are engaged on task are three important characteristics of a skills development program that enhances student learning outcomes. This is also true of making explicit the criteria on which students' work will be judged and making explicit the requirements for students' self-regulation of their own behaviour (Lingard, Mills and Hayes, 2000). The construction of a skills development program framed to support productive linguistic diversity involves providing a supportive teaching and learning environment. This includes engaging with the complexities of the teaching/learning environment; recognising the fragility of teaching/learning environments; making provision for maintenance of existing teaching/learning practices during the processes of innovations, and overcoming discontinuities in resource provision.

A supportive skills development program for Australian and international students who speak a language other than English, is one that enhances students' proficiency in both languages. Thus, in this context it is not a matter of choosing between promoting the sustainability of global multilingualism or advancing the project of globalising English, a project in which the Oxford English Dictionary played a key role (Winchester, 1998). Second-generation digital technologies are making such a choice redundant and now educators are exploring possibilities for creating skills development programs that are supportive of the extension of students' vernacular. In doing so it may even be possible to retard the projected extinction of 90% of languages by the end of the twenty first century (Crystal, 2000). Of the range of factors important to the sustainability of humanity's multilingual knowledge base, two are particularly relevant here:

- Firstly, there must be a willingness among those in the business of globalising the English language (and other global languages), to commit themselves, each in their own small way, to this ethical investment project.
- Secondly, there must be a sustained commitment to and support for the inter-generational transmission of these languages. This can be facilitated through multilingual book production.

With the beginning of the new millennium there has come a marked shift to the use of second-generation, multilingual digital technologies. This has seen a shift away from the project of globalising English towards the rising production of electronic and digital publications in national languages and increasingly in local languages. There has also been a recognition of the creative development of different and distinctive varieties of English, which have produced grammatical, lexical and stylistic innovations (Gupta, 2001). Recognising these trends, some businesses now acknowledge that productive engagement with linguistic diversity is integral to their positioning within the global multilingual knowledge economy. For instance, this shift to multilingualism has found expression in the strategy of 'globalised localisation':

> ...corporations seek to maximise their market share by shaping their products to local conditions. Thus, while CNN and MTV originally broadcast around the world in English, they are now producing editions in Hindi, Spanish, and other languages in order to compete with international and regional media outlets (Warschauer, 2000, p. 157).

Language revitalisation projects are not concerned with creating a static linguistic archive that records an important part of people's past, but with 'bringing the language forward' so that it is part of people's future (Warschauer, 2000, pp. 166–167). Innovation in the electronic and digital creation and consumption of books can be part of these language revitalisation projects helping to bring the world's languages forward so that they are part of people's future. However, it is important to recognise that:

> ...language is not a self-sustaining entity. It can only exist where there is a community to speak and transmit it. A community of people can exist only where there is a viable environment for them to live in, and a means of making a living. Where communities cannot thrive, their languages are in danger. When languages lose their speakers, they die (Nettle, and Romaine, 2000, p. 5).

WHAT PROCESSES ARE CONTRIBUTING TO THE PUBLIC EDUCATION OF INVESTORS ABOUT ISSUES OF CORPORATE RESPONSIBILITY?

Investors are being educated about the requirements for corporate responsibility through accounts provided in the mass media of people's responses to the social, economic and cultural impact of various initiatives to further the interests of 'globalisation from above' (Falk, 1999). For instance, there have been direct action campaigns against companies involved in uranium mining, or associated with apartheid, as well as campaigns against the proposal to establish a Multilateral Agreement on Investment that would have made foreign companies immune from national laws. As a result of these protests, S11 and M1 emerged raising further concerns against the economic irrationalism of what is seen as not-quite-free trade. Media reports of demonstrations, blockades and even shareholder activism are part of the public education of investors in what constitutes corporate social and environmental responsibility.

Ethical investment addresses shareholders' concerns about the future of the planet, the needs of future generations, and the future of humanity's linguistic diversity and associated knowledge base.[2] How will the Australian publishing and printing industry respond to investors who actively screen out those companies ethically at odds with their commitment to the sustainability of bio-linguistic diversity?

Languages are the archives of human knowledge, giving expression to events that formed people's past, as well as providing the intellectual resources for making judgments about present actions and future developments. The loss of one's language means the loss of that knowledge and thus the exclusion of that knowledge from considerations about the present and future (Crystal, 2000, pp. 40–43). Of the 6000 languages in the world, only 600 are expected to safely survive the course of this century. After several millennia, most the world's multilingual knowledge base is in danger of extinction within the next one hundred years (Nettle and Romaine, 2000). The loss of any language means the loss of knowledge about human organization, the environment, categories of experience, culture, technology, medicine, climate, art, music, history and much more.

---

[2] www.ethicalinvestor.com.au; www.eia.org.au

In this era of globalisation the loss of this knowledge means the loss of a potentially valuable economic, scientific or even military resource. At the very least we know that to date the loss of this knowledge has been accompanied by the over-exploitation and degradation of the land and sea. Nettle and Romaine (2000, p. 16) argue that there are '... many striking similarities between the loss of linguistic diversity and the loss of biodiversity... The areas [of the world] with the greatest biological diversity also have the greatest linguistic/cultural diversity.'

Increasingly, financial planners, including those with responsibility for management superannuation funds, are becoming aware of ethical investment, often as a risk management strategy. There are also shareholders who seek to align their needs for financial security with their commitments to the values of social justice and environmental sustainability. James Rose (2001, pp. 66–67) the CEO of Integrative Strategies provides a list of ethical shareholder concerns which, when applied specifically to the printing and publishing industry, raise questions for investors such as the following:

- How is the industry contributing to environmental sustainability, including the knowledge available in different languages, with regard to this concern?
- What contribution is the industry making to human rights, including the inter-generational transmission of languages?
- What is the global contribution of the industry to the protection of Indigenous rights, including the bringing forward of their languages?
- How is the industry interacting with and integrating itself into transnational, multilingual communities?
- How is the industry contributing to efforts to prepare and maintain a multilingual workforce?
- What is the industry's policy on investing in the multilingual knowledge economy?
- What is the relation between the industry's public relation campaigns and its actual substantive contribution to the sustainability of linguistic diversity?

CONCLUSION

This chapter has provided an overview of issues relating to the use of second-generation digital technologies in multilingual book production for the multilingual knowledge economy. The multicultural and multilingual features of Australian industry and its workforce have been briefly outlined. A skill development framework to support the skilling of Australia's multicultural and multilingual workforce has been described in terms of the key features of quality skills development programs.

Contemporary manifestations of the historical phenomenon of globalisation are evident in the internationalisation of the economy and worldwide integration through second-generation digital technology. In cyberspace, as in other domains of this new technology, issues of multiculturalism and multilingualism matter as much as ever. Multilingualism is a fast expanding field for those electronic and digital businesses seriously interested in carving out a niche in the restructured international marketplace. There are opportunities for the printing and publishing industry to position itself within the global multilingual knowledge economy and to support the maintenance of the world's multilingualism through the use of second-generation digital technology. Professional translators and interpreters can be trained to play an important role in the industry as language brokers and cultural mediators. Bilingualism, or proficiency in a second language, can provide added value to Australian industries seeking to secure a competitive advantage in the multilingual knowledge economy. For Australian governments keen for its industries to participate in the multilingual knowledge economy investment in the education and training of a globally oriented workforce with diverse linguistic skills is as necessary as it is essential.

Preparing for a globally-oriented, multilingual publishing and printing industry calls for concerted investment from the public sector and industry to produce and use multilingual services, tools and systems.

REFERENCES

The Age (2001), Start-up translates idea into language of success. June 12, p. 9.

Ashenden, D. and Milligan, S. (2001), Mind your languages, *Education Age*, Wednesday April 4, pp. 8–9.

Auh, T-S. (1999), Promoting multilingualism on the Internet: Korean Experience. (http://www.unesco.org/ webworldinfoethics2/eng/papers/paper_8.htm accessed March 5 2001).

Australian Bureau of Statistics. Population languages. (http://www.abs.gov/ ausstats/ABS@.nsf/).